# Coppicing
# & Coppice Crafts

WOOD

# Coppicing & Coppice Crafts

## A Comprehensive Guide

Rebecca Oaks and Edward Mills
Foreword by Professor Oliver Rackham

The Crowood Press

First published in 2010 by
The Crowood Press Ltd
Ramsbury, Marlborough
Wiltshire SN8 2HR

**www.crowood.com**

This impression 2012



**British Library Cataloguing-in-Publication Data**
A catalogue record for this book is available from the British Library.

ISBN 978 1 84797 212 5

**Disclaimer**
Chainsaws and many other tools used in coppicing and the coppice trades are potentially dangerous. All chainsaws and other tools and equipment used in coppicing work should be used in strict accordance with both current health and safety regulations and the manufacturer's instructions. The authors and the publisher do not accept any responsibility in any manner whatsoever for any error or omission, or any loss, damage, injury, adverse outcome, or liability of any kind incurred as a result of the use of the information contained in this book, or reliance upon it. If in doubt about any aspect of coppicing or the coppicing trades readers are advised to seek professional advice.

Designed and typeset by Focus Publishing, Sevenoaks, Kent

Printed and bound in Malaysia by Times Offset Malaysia Sdn Bhd

# Contents

# Acknowledgements

RO

Thanks to all those who have supported me whilst writing this book. My colleagues and trusted apprentices, Mike Carswell and Sam Ansell, who have kept the business going in my distracted absence. Friends who have proof-read the chapters: Pat Urry, Lynne Alexander, Helen Shacklady and Kath Morgan. All my colleagues involved in the Bill Hogarth MBE Memorial Apprenticeship Trust and Coppice Association North West for their dedication to the cause and constant reaffirmation of the importance of coppicing. Particular thanks to Brian Crawley, Paul Girling, Edward Acland, Twiggy and, further afield, Steve Homewood and Alan Waters. To Bill Hogarth himself who inspired and mentored me from the start. Thanks to Julian Bingley for inspiration for the map and to Madeline Holloway for lending me a pen. Lastly, thanks to my partner Amanda Bingley for her unstinting support.

EJM

I would like to thank my employers and their funders for allowing me to work on this book during time for which they were partly paying; they include Cumbria County Council, the Forestry Commission and Natural England. Thanks also to my colleagues for their support while I wasn't in the office. I offer my thanks to the numerous people who have helped with advice, comments and images, including (in no particular order) Chris Starr, Alan Shepley, Alan Waters, Debbie Bartlett, Jackie Dunne, Iris Glimmerveen, Tim Youngs and Teresa Morris for her fantastic photographs of deer. I would like to thank my family for being patient while I have been tucked away in my study and not attending as much as I should to family and household affairs. I owe a debt of gratitude for the inspiration of Oliver Rackham, Bill Hogarth, Colin Simpson and Martin Clark.

We should also like to thank the publisher, The Crowood Press for their advice and support.

All photographs are by Edward Mills and Rebecca Oaks, except where indicated.

# Dedication

This book is dedicated to the memory of Colin Simpson.

# Foreword

Coppicing makes use of a mysterious property that most trees have: when cut down they do not die but grow again from the stump or roots. People have used this behaviour for at least 6,000 years to generate renewable supplies of wood for fuel or to use for many crafts, simple or specialized. For centuries most woods in England and Wales were coppices: woodland plants and animals became adapted to cycles of years of light and years of shade. In the twentieth century coppicing fell into decline owing to competition from fossil fuels, competition for labour, competition for land from the rising fashion for plantation forestry, and lack of organization and marketing. Woods were neglected and became continuously shaded, to which their plants and animals were not adapted. Since the 1950s interest in coppicing has revived among those interested in conserving woodland ecology, followed by those interested in woodland crafts and in reviving renewable energy. At the same time it has encountered a new threat from ever-increasing numbers of deer, not present before, which devour the young shoots.

This book is written for anyone interested in coppicing – for anyone acquiring a wood and wanting to know what they are letting themselves in for, for those seeking to understand what the county wildlife trust is doing to its woodland, those who love the nightingales and the ancient many-stemmed trees and the spring flowers, and anyone who delights in making things.

Professor Oliver Rackham, Corpus Christi College, Cambridge

# Chapter 1
# Setting the Scene – An Introduction to Coppicing

Sustainability is a word that has been much overused during the last twenty years. However, the art of coppicing must be the ultimate sustainable woodland management technique. In a well-managed coppice there is no need for herbicides or fertilizers, no need to disturb the soil or to plant trees, no need to use tree shelters or to put up fences. The coppice is entirely self-renewing. Those who have coppiced trees and experienced the magical transformation of regrowth springing from the cut stump will never forget it or cease to be amazed by it each year. The burst of new life stimulated by the sun's warmth and light is surely one of Britain's most under-rated wildlife wonders.

As the coppice grows, a revolution occurs. Over one growing season the bare, forlorn coppice floor becomes a scene of riotous new growth, burgeoning shoots growing at a rate that you can almost see. In the case of the ultimate coppice species, hazel, in a few short years the rods are ready to be cut again and the quality of the raw materials will be evident. These self-renewing, sustainable materials are used to make products as varied as thatching spars to bean-poles and hurdle rods, and seem to be growing if not for free, then simply for the effort you put in to coppicing them.

*Burgeoning life in a newly cut coppice.*

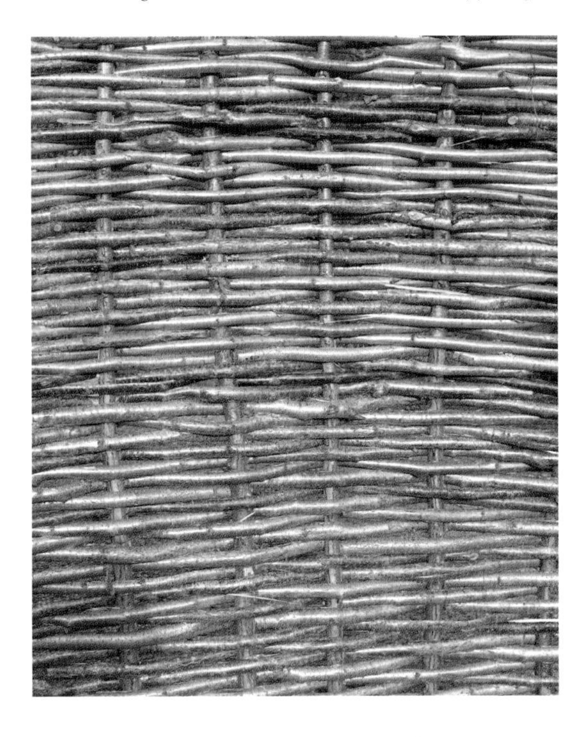

*Wattling; strong, light and very versatile.*
*(Photo: Anna Gray)*

*Seven-year-old hazel ready to be cut again.*

The knowledge of our ancestors of the potential of woods to renew themselves with speed and abundance is equalled by their ingenuity for finding uses for this natural bounty. We can only speculate how far back the skills of the hazel hurdle-maker reach, but we do know that the wattle of our old timber houses must have a lineage that stretches from our earliest attempts to fashion a shelter from entwined branches. The transitory nature of the material leaves no trace in the archaeological record, except of course in those fabled Somerset peat bogs, where trackways dating back to 3000BC have been found, made of 'interwoven underwood rods, grown for the purpose' (Rackham 1986). So from houses, to roads, to fences and pens; all manner of necessities for human survival were first sprung from the coppice woods of Britain.

Amazingly, although coppice woods have been out of favour or out of fashion for at least fifty years, we do still have a coppice industry in Britain. Coppice workers have weathered the advent of plastics that have made our rich and varied heritage of baskets almost obsolete, and sent a simple wooden item such as a cotton reel to oblivion. Post-World War II, a chemical revolution almost put paid to the oak bark tanning industry and a tide of cheap plastic imports replaced many of our coppice crafts. This mass manufacturing threatened to chase all our home-grown craftspeople out of the woods; however, an appreciation of natural products and skills has saved these stalwarts from extinction.

At a recent coppice conference held in the North-West of England, the range of experience of coppicing was striking. There is a remarkable variation in what coppicing means to people today. In effect, there is a time-line that stretches the length of the country starting in Kent, where the chestnut coppice industry is still fairly intact, and leading up through the country to the Highlands and islands of Scotland, which have not seen coppicing for many generations. Kent and East Sussex are fortunate to have a chestnut paling industry that employs people who have a family connection with coppicing. These people often started their coppicing career straight after leaving school and took up specialist skills within their industry, following directly in the footsteps of their forebears.

*The old and the new: swill baskets and plastic bags.*

As you move west you come into the heartland of the hazel coppice industry in West Sussex and Hampshire, where the evidence for the recent coppice industry is plain to see in the swathes of coppiced woods in a variety of conditions from active management to dereliction. Hazel coppice, in particular the hurdle industry, went into decline when these products were out of favour and the markets looked like they were disappearing. Jonathon Howe highlighted the decline of the hazel coppice industry in his report to Hampshire County Council (1991) and stimulated a fresh look at the potential of this resource. Today there is a healthy number of folk involved in coppicing in Hampshire, both those with a long-term family connection to coppicing and a fair scattering of 'newcomers' who have carved a more modern model of coppice working. At the NW coppice conference, Professor Ted Collins drew our attention to these two differing styles of coppice worker: the ones who adhere to an 'old tradition' of specialist skills, the 'hurdlemaker', 'spar maker' and 'woodcutter', who joined the industry straight from school and have years of experience and knowledge to draw on; and a new breed of highly educated coppice workers who have mixed businesses with an emphasis on teaching and public education, 'spreading the word' like modern coppice evangelists (Collins 2004).

These modern coppice workers can be found in all regions that have a vestige of recent coppice history. In the Midlands and the East of England, and even in the South-West, continuity of coppicing was broken only a generation ago. If you are lucky there may be a few retired coppicemen to pester for information on how, where, when and why coppicing was done. In South Cumbria, North-West England, Bill Hogarth MBE (1929–99) worked as a coppice merchant. He came from the 'old tradition', learning his craft from his father. Bill knew all about the long hours and high productivity that was essential to make a living producing, what were essentially low-value products, such as bobbin wood and ships' fenders. Thankfully he also realized the importance of passing on his skills to a new generation, and was always generous with his knowledge.

*A chestnut paling mountain.*

*Alan Waters demonstrating at a local show.*

*Bill Hogarth engaging a group of students.*

Further north again, in amongst the fabulous oak/hazel woods of Argyll, the evidence for coppicing is found in historical records and woodland archaeology. It is much harder to reinstate coppicing when there is no one left to remember and to argue the case for this form of woodland management.

The evidence for coppicing remains in many woodlands, often in the beguiling guise of our woodland flora. It is no great mystery that the demise of our woodland butterflies has gone hand in hand with the loss of coppiced woodland. The great resurgence of coppicing as the woodland management tool of choice has been driven in many parts of the country not by economics but by conservation. Habitat restoration and species protection have stimulated an explosion of spending on coppice restoration in a few areas. But in the past we didn't have to rely on the public purse to create the habitat that these creatures need: the coppice workers throughout history did that as an entirely incidental by-product of their industry. Today, coppice workers are endeavouring to bring these two sometimes opposing approaches together to create habitat- and species'-rich woodlands, at the same time as carving a living within the woods.

Although coppicing as a woodland-management technique is, in all probability, thousands of years old, it is as relevant today as it has ever been. Coppice workers are skilled and passionate about delivering all the 'outputs' that we demand from our woodlands. They are the epitome of the sustainable rural livelihood, living close to their natural resource and supplying basic requirements, such as firewood, as well as products that may be considered a luxury or just simply beautiful. By and large they are a contented group, due to the healthy outdoor lifestyle and the satisfaction of working with nature, who get immense satisfaction from making and selling products that may seem obscure.

*Edward Acland making ash gate hurdles.*

It is essential that the skills of coppicing and the coppice crafts are kept alive today and into the future. If all our old coppices are converted to 'non-intervention' woods, as some would want us to do, there will be a net loss of biodiversity. Many people think that nature can look after itself and should be left to do so. However, humans have been coppicing, quite possibly in some places for thousands of years, and even if wildlife has not actually evolved to take advantage of this, some wildlife has certainly taken advantage and greatly enlarged its population. In fact it is hard to believe that some wildlife, which has a short annual life-cycle, has not evolved over thousands of generations to maximize the use of coppice rotations to enhance its success as a species. There are some experts who would have us believe that coppicing should not be contemplated for fear of a loss of biodiversity, even though there is clear evidence that it has been carried out in the recent past. Only a small fraction of the area coppiced, say, one hundred years ago, is still coppiced today; in contrast, there are thousands of hectares of rather bland, uniform woods that will take another thousand years to capture the wildlife of an old coppice.

There has been much interest in the twenty-first century in education in a forest setting – the Forest Schools or Woodland Classrooms concept. This is a tremendous idea, is rightly very popular and works very well in a world where children are increasingly disconnected from their natural environment. One of the authors is convinced his passion for trees and woodlands was begun by a visit to a school wild area, which was in fact a medieval hedgerow and was packed full of an amazing variety of species in a small space. So, what better place to study than an actively managed coppice? Children can not only learn the usual standard numeracy and literacy, but can also cut materials and make things from them. They can learn how to build dens, make fire and to use it for cooking. Many children will remember this kind of experience for the rest of their lives and, for some, the experience will be a life-changing one.

In a similar way, coppicing by hand is a fantastic way to get volunteers involved in the great outdoors. People can learn about the wildlife of coppiced woodlands and observe it at first-hand, as well as taking home raw materials or simple products they might have made. It's a healthy pursuit and a first-time coppice volunteer will find that muscles have been used that have not been used for a long time.

Some people who have tried their hand at coppicing as a volunteer discover that they have a creative streak and have a hidden talent at making things with the materials they've cut. Clearly this is important and should be encouraged as there is a legacy in Britain of crafts and skills, many of which have already been lost or forgotten. There is a need for continuity to preserve and honour this heritage of coppice skills and crafts.

This book aims to show you how to fulfil your dreams; whether you are an aspiring land-owner, an enthusiastic volunteer or a budding coppice worker, to acquire a wood, get the equipment and the skills that you need, and to start coppicing. We hope to inspire you with the chapters on the wildlife benefits and the fantastic products that can be gained from the woods. And, finally, to encourage you to take the plunge, give up the day job and make coppicing your life.

*The woodland classroom: inspiring the young.*

# Chapter 2
# Looking for a Suitable Wood

## INTRODUCTION – WHAT ARE YOUR OPTIONS?

Given the option, most people who have ever even thought about it, would want to own their own woodland. Of course, this is frequently just not a practical proposition given the value of land. However, since the turn of the century, woodland ownership has become more popular and much more accessible, despite the cost.

Although there is a lot of information in this chapter, and it's full of 'make sure' and 'don't forget to', it's actually incredibly easy to purchase a wood and wonderfully satisfying. Please don't let any of the guidance in this chapter put you off. After the purchase is made and you visit your new acquisition, the concept of woodland ownership can seem unreal. For a novice land-owner, to be able to walk through your own woods and realize that it really is up to you how this wood is managed, can take a long time to sink in.

You may want to consider purchasing your wood with family and/or friends. This could give you a larger budget to look for a bigger wood, and more resources when it comes to managing and looking after it. It can also be more fun to share the work and the enjoyment with other people. If buying a wood is beyond your means, then do not despair as there are many other options such as leasing or short-term contracts, all of which we will look at in this chapter. The advice given here is relevant regardless of whether you own or lease the wood.

## FIRST, FIND A WOOD

Woodland purchase might seem daunting, but the process is just like purchasing any other land-related property, such as a house. The most common difficulty is how to find a wood on which to make an offer. Most housing estate agents don't deal with land. This is usually the preserve of more specialist companies such as land agents or auction marts. The best place to keep an eye on what's coming up for sale in your own area is the local newspaper, especially if it's a rural one with a farming section. Once you can identify which companies normally deal with land sales, you can keep checking their website too. Although word-of-mouth is a valuable way of finding out which land is coming up for sale, it can be difficult to get into this network unless you already live and/or work within that rural community. If your heart is set on owning a wood regardless of where it is, you'll need to rely on a company that has national coverage, and keep checking their website or ask to be put on their mailing list.

### *Check out the Legal Issues*

It is just as important to get your solicitor to carry out the local authority searches on a woodland purchase as it is on a house. There could be issues about planning, access or water supply, for example. It is essential that you read the particulars of a sale carefully and, if you have any queries or if something isn't mentioned, then you or your solicitor should ask. On the opposite page there is a checklist of things to find out.

### *What to Pay?*

There are a number of features that can either enhance or lower the value of a wood on the open market. Features that can increase the amount that people are prepared to pay include:

**Checklist of things to check when you purchase woodland**

- The exact boundaries of the land.
- Who has responsibility for the maintenance of the boundaries? Although it might be attractive if neighbouring land-owners are responsible for all the boundary hedges, walls or fences, how easy will it be for you to persuade them to maintain them? It may be easier for you to have the responsibility.
- Access to and from the land – if you don't own direct access off the public highway, you will need a formal (legal) right of access across someone else's land.
- Is water supplied to a neighbour from the wood? This is usually only an issue in upland areas; it shouldn't affect the way you manage your wood.
- All woods have shooting rights attached to them. You need to check whether these have been sold separately at some time in the past or whether they still go with the wood and are, therefore, included in the sale. If you do not hold the shooting rights, the person who does is entitled to shoot game or pest species in your wood, whether you approve or not.
- Are the mineral rights included in the sale? If you don't hold the mineral rights (a relatively unusual situation), they are probably held by a large quarrying or mining company. Clearly, if a wood is quarried, there won't be much left of it at the end of the quarrying operation!
- Are there any special covenants on the land? Land is sometimes sold with a covenant prohibiting erection of buildings, for example.
- Are there any grant schemes on the land that are current? Some grant schemes have liabilities attached to them; for example, if you purchase a piece of land with a 're-stocking obligation' on it (i.e. the owner is required by the Forestry Commission to replant trees after a felling operation), as the new owner you may be obliged to fulfil it.
- Is the land located in a designated landscape; for example, an Area of Outstanding Natural Beauty or a National Park?
- Does the wood itself have any designations? If the wood is designated as a Site of Special Scientific Interest (SSSI), you will need permission to do almost anything in the wood. If it is covered by a Tree Preservation Order or it is within a conservation area, different rules apply – see Chapter 3.
- Are there any services, wayleaves or easements in the wood? These could include water and gas pipes, and electricity and telephone cables, either above or below ground. Utilities companies will normally have a right of access across your land for maintenance purposes.
- Liabilities and hazards. If the wood encompasses a road or other rights of way, there will be health and safety issues that will have to be addressed. Usually, the greater the hazard, i.e. if lots of people use the right of way, the more the risk needs to be reduced. Although this kind of thing shouldn't stop you purchasing, leasing or managing a wood, it must be considered. If there is tree safety work needed, it can be expensive and you will have to take responsibility for getting the work done.
- Does anyone have third-party rights? The neighbouring farmer may have a right to take livestock through your wood from one field to another; this could involve several tractor journeys per day.

good access to and within the wood; woodlands located in well-known areas of landscape beauty; timber of commercial value; a water feature, such as a pond, lake or riverside; or fishing rights. Usually, woodland in the South-East of England is valued at more than woodland elsewhere in the country. Clearly, broad-leaved or mixed woodland with bluebells in spring will fetch more than a dark spruce plantation. Woodlands that are lower in value include those that are remote with poor access; younger, less mature woods with a lower commercial timber value; conifer woodlands generally, and especially those that have been unthinned or are neglected. Woods with apparent safety hazards and liabilities will also fetch a lower price.

*Woods with water usually fetch a higher price than those without.*

*Conifer woodlands generally cost less than broadleaved woods.*

*A re-stock site involves a lot of work but may be relatively cheap.* (Photo: Ted Wilson)

The price of woodland and forests is usually referred to per acre or hectare (ha). For the new woodland owner this may be helpful as a rough guide, but sometimes the value of the wood to a prospective buyer bears little relation to the traditional price per hectare of woodland. Some buyers may be quite happy paying the same as a large new car for just an acre or two, whilst others search for better value for money. As we write, broad-leaved woodland in England rarely sells for less than £3,000/ha (£1,400/acre) and can sell for up to £25,000/ha (£10,000/acre); normally, there is an economy of scale in purchasing larger woods – these sell for less per hectare, whereas smaller woods sell for more per hectare.

If you find a wood that is coming up for sale by auction and you've never been to a land auction before, it is worth going to one to find out how it operates and to familiarize yourself with the situation. If you have any questions or there is anything unclear about access or shooting rights, for example, do make sure you get clarification on these vital issues before bidding begins. Auctions can be very exiting (especially if you are bidding!) and often have a tense atmosphere that can be cut with a knife – groups of people whispering suspiciously, furtive glances to competing parties – it can all be rather intimidating. If you've decided to bid, be sure what your upper bidding limit is – once the hammer goes down on your winning bid, you're legally obliged to purchase. Remember that before you leave the auction room, most auctioneers require a deposit of 10 per cent (usually), so you'll need your cheque book and the available funds in your bank.

## Wood-lotting

There are companies now that specialize in purchasing woods and dividing them into smaller and thus cheaper plots, making purchase more affordable. For some people (and some woods), this is an excellent solution, putting woodlands in need of sympathetic management in touch with people who have the resources and enthusiasm to invest in improving the wood. However, sometimes these companies have come under great criticism for purchasing large woods and sub-dividing them into many very small plots (some of the plots are as small as 30m2) and selling these for immense profits. This has become known as 'wood-lotting' and has the potential to be a disaster for the wood.

If you are tempted to purchase a wood-lot, even if it's not that small, you need to think very carefully before spending your hard-earned money in this way. If a wood is divided up into a hundred units, it is unlikely that their owners will all be working to the same management objectives. A study has shown that 75 per cent of wood-lots are in ancient semi-natural woodland (Kent Wood-Lotting, Land Use Consultants 2007). Some owners purchase their wood-lot as an investment or with the aim of obtaining planning permission, with no intention of managing the wood. Most do not obtain advice and are often unaware of statutory controls such as felling licences or tree preservation orders (although at least one of these companies provides free membership of the Small Woods Association and the Royal Forestry Society as part of the sale and currently has a free website packed full of resources). Some owners erect tool sheds or other structures, which results in a net loss of irreplaceable ancient woodland, and some lots are incorporated directly into back gardens where the lot abuts an urban area; some owners keep horses or pigs, causing damage to the wood and there are sometimes damaging recreational activities.

All in all, the story of wood-lotting is not a very happy one and frequently results in a net decline in the condition of the woodland.

## Renting or Leasing a Wood

You might decide that leasing or renting some woodland is a better option than purchase. Of course, it's certainly a cheaper option, but it may well be as difficult to find a wood to lease as it is to buy. Woods to rent are only very rarely advertised, despite there being thousands of hectares of neglected woodland that would benefit from the kind of care that many people could give it. The only probable way of finding a woodland owner likely to enter into this kind of arrangement is by word of mouth and personal approaches.

Some owners would expect to be paid an annual rent but there are other kinds of owners who may allow you to manage woodland in return for your services; this could be supplying them with firewood, or perhaps renewing a fence or hedgerow boundary, improving access to the wood or some other return from your activity in the wood. It may be worth offering a return of this sort to the owner as any cash rental income from the wood for them is likely to be relatively small (even if it might be large for you!).

It is important to have a contract to protect both parties. This should stipulate when the rent or lease period starts and finishes, what the payment is (if any), any times when the owner does not want access (e.g. during a shoot), who is responsible for the repair of access and so on. In the same way that an owner should have a contract for any work going on in a wood, or for timber sales, you need to make sure that you do not find yourself liable for unexpected costs. The contract should also include what you are going to do in the wood and who is responsible for applying for the felling licence and any other permissions needed (see Chapter 3 for more on felling licence regulations and other legislation).

Some owners will feel more comfortable if you can provide references and show them evi-

*Using a chainsaw with all the Personal Protective Clothing (PPE).*

dence that you know what you're doing! Certificates from relevant courses attended could be helpful; if you intend to use a chainsaw, you will definitely need to hold the NPTC CS30/31 chainsaw certificate in order to satisfy your insurance company and that of the owner. This is a grey area, and many owners will be happy to allow you to work in their woodland without insurance, risk assessments or the relevant qualifications. This is all well and good until there is an accident and then problems arise with liability, compensation and the potential for a very expensive court case. If you negotiate an arrangement with an owner whose woodland is entered into one of the certification schemes, you will definitely require all the relevant certificates. This is likely to include a current chainsaw certificate and public liability insurance as a minimum.

## READING THE WOODLAND

### First – Look at Maps of the Wood

If a woodland is coming up for sale that you think might be suitable, first look at the woodland on a map. The spacing of the contours will give you a good idea of the terrain before you visit. You will also be able to tell from the map which way the wood faces (aspect); although this is probably not important for most people, you could be disappointed if you have your heart set on evening picnics in the sunshine, but the wood faces east and is in the shade all afternoon. The map will also tell you the altitude of the wood; this, along with aspect and exposure, will give you an idea of the risk of windblow (see Chapter 4 for more on windblow).

There are other things you can find out by looking at a map. Is the wood isolated in farmland or is it part of a more extensive woodland network? Is there anything of interest marked on the map within the wood? Is there a working quarry next door or other potentially noisy neighbours? There is a wide variety of features that a map can give you clues about before you make your first site visit. These can either put you right off the wood straight away, or give you further encouragement to visit. They can include old buildings, earthworks, rights of way, tracks, ponds, wayleaves (including pylons), industrial works and so on.

### Second – Visit the Wood

Even if on paper the wood looks ideal, it really makes sense to visit it. This sounds like an obvious piece of advice, but you'd be surprised! There is no substitute for getting the 'feel' of a place. Your first impressions will be invaluable and you can gather some basic information in a very short time. Is there a shrub layer? What is the age range of the trees and is there any natural regeneration? Is the wood obviously going to present some challenges – is it steep, wet, rocky, down a steep, narrow lane, on top of a hill or next to the local pig farm? Perhaps you'll fall in love with the place after a quick visit telling you all you need to know to encourage you to make an offer – wonderful bluebells, beautiful hazel coppice, the song of a nightingale or a view that you could look at forever.

### Topography

The disappointment of a wood that faces the wrong way has already been mentioned,

*Windblown oak and birch.*

*Your wood may be an important landscape feature.*

although for most people, this won't be too important. The main impact that the land form has on the woodland is on the ability to manage it. Steep and rough ground can affect access routes – tracks and paths – especially ease of timber extraction; it can affect the likelihood of windblow; the land form and aspect can have profound effects on wildlife. A north and east-facing woodland will be cooler and damper than a south and west-facing wood, but it will also be more sheltered and less susceptible to windblow.

As well as the view from the wood, the topography can affect the view of the wood from the wider countryside. This is usually not a factor you need to consider in flat countryside, but in hilly areas and especially in designated landscapes like national parks and areas of outstanding natural beauty, what the woodland looks like in the landscape is important. Foresters still receive criticism for the straight lines that new forestry plantations created in the 1970s, although in reality there are now very few of these remaining. A poorly designed coppice coupe or access track could become a regrettable landscape eyesore.

### Access

Good access is invaluable but the level of access you need depends very much on your objectives. If you are not expecting to drive into the wood in a car, then a rough track is fine. If there is never going to be large timber to bring out of the wood, then access can be relatively simple. If, however, you are planning on coppicing and bringing out a variety of produce, basic access is extremely important. You will soon get tired of carrying bits of wood any distance – you might be able to manage this now, but in twenty years? A bundle of twenty-five freshly cut hazel rods is very heavy and, unless you are super fit, you won't be able to carry it very far.

Proper, formal access tracks made to a good standard are expensive to create. There are shortcuts, but there will be usually be considerable costs, whether this is purchase of bought-in materials, equipment hire, payment of contractors or simply your own time. Therefore, if your wood already has access, this will save you a great deal of time and expense, and this is a key thing to look out for when searching for a wood to look after and enjoy.

*Good access is critically important.*

Access within a wood may be fine, but sometimes, access to the wood can be difficult. This is especially so in very rural areas and in hilly countryside, where lanes can be steep, narrow and deep. This can make getting timber away from a wood difficult and sometimes impossible. Although you can improve access within a wood, it is almost impossible to improve access to a wood.

You may not want a car park within or next to your wood, but many woodland owners will end up having to drive to their wood and so, you will need somewhere to leave your vehicle. If there is no access track, you may need to negotiate with a neighbouring land-owner, so you can leave a vehicle in a farm yard or field gateway from time to time. If you ever have friends to visit the wood, or work parties, you may need space for several vehicles.

Many woodlands have one or more public rights of way (footpaths, bridleways or byways open to all traffic – BOATs) running through them. Some people cherish their privacy and wouldn't consider a wood with a public footpath in it; for others, this is not important. However, it is certainly the case that having the public in your wood can be an issue. The main area of concern is the security of any equipment and materials that are stored in the wood. Ideally, equipment would be stored somewhere secure, away from the woodland, but this is not always practical. People will wander off public footpaths – most people aren't very good at map-reading – although with good waymarking and other management techniques that deter people from leaving the path, this can to an extent, be managed. Whether you have public access or not, the wood will need to be covered by public liability insurance.

### *Boundaries*

The condition of the boundaries is important, especially in the uplands and other areas where there might be grazing livestock in adjacent fields. Sheep will eat coppice re-growth and need to be kept out of most woodland at all costs. Cows are generally less destructive and much easier to keep out. However, in just a brief period at high densities in wet weather, cows can decimate ground flora with their heavy weight.

Historically and legally, a famer is obliged to keep livestock in his or her own fields – it is not up to a woodland owner to keep someone else's livestock out of his wood. However, through a subtle change in countryside 'lore', you will often hear this repeated the other way around! A common phrase heard when discussing boundaries these days is 'a wood fences itself' meaning that the woodland owner is responsible for keeping livestock out. Legally, this is nonsense and you can argue this with authority. Sometimes, if a land-owner is selling a wood and retaining land surrounding it, the boundary maintenance obligations may be sold with the wood. You will need to foster a positive working relationship with your woodland neighbours, and boundary disagreements are likely to be a source of dispute unless both you and your neighbour are prepared to compromise and share costs.

## DECIDE ON YOUR OBJECTIVES

We will assume that, as you have picked up this book, your main aim is to find a woodland in which to carry out some coppicing. You need to decide what you want to do with your wood and what you want your wood to do for you. So, before you launch into purchasing, leasing or entering into a management agreement, identify whether coppicing is both possible and desirable.

*You should check whether boundary maintenance is your responsibility. This fence was replaced to keep out sheep, about five years before the photo was taken.*

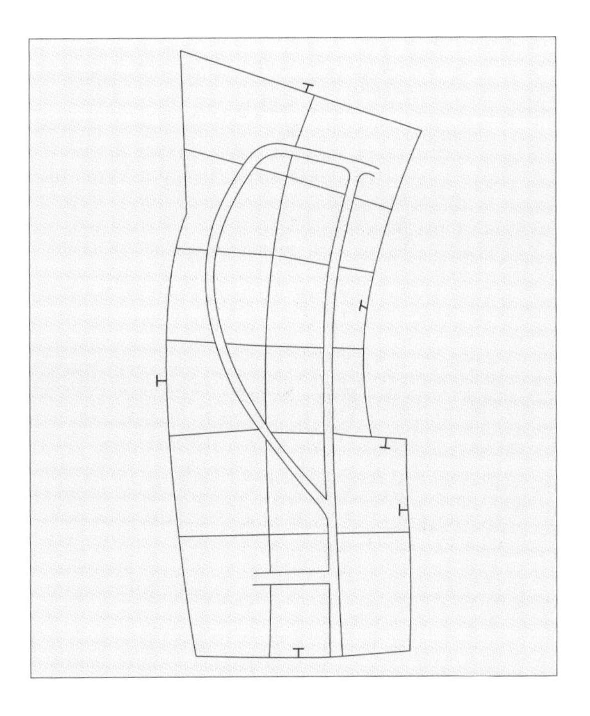

*Boundary marks – if the T is facing into the wood, the boundary is yours; if it faces onto adjoining land, the responsibility for boundary up-keep is your neighbour's.*

### *Historical Precedence*

Woods that don't have a history of coppicing are unlikely to make successful coppices now, without a lot of work. The main exception to this rule of thumb is for some recently planted woods.

Ideally then, for your prospective wood to work as coppice, it will show signs of past coppicing. This means finding lots of multi-stemmed trees. Sometimes evidence on the ground can be supported by looking at early edition Ordnance Survey maps that actually carry a mark that signifies coppice. Seeing multi-stemmed hazel, however, is not necessarily a sign of coppicing. Hazel naturally grows with multiple stems. Even in the furthest reaches of the most remote wooded gorge you could possibly imagine, hazel produces multiple stems and appears to have been coppiced; this is its natural growth habit and is not necessarily a sign of past human activity.

If you can find other multi-stemmed trees, they should ideally, not be older than about fifty years. The longer a coppice stool remains uncut, the less likely it is that it will survive being coppiced again. This is particularly the case for oak and birch which respond poorly if not coppiced for several decades. Better results are obtained with older hazel, ash, sweet chestnut, field maple and alder, for example. If the coppice stools are a long way apart, in order to get a productive, working coppice, new stools will need to be established in the gaps. This obviously creates work and expense that you may not want; conversely, it could give you the opportunity to influence the composition of the wood and give you more of the coppice species you want.

*High forest broad-leaved woodland with virtually no potential for coppice.*

There are some other tell-tale signs that a wood might have been coppiced in its past. In some areas, there are the remains of woodland industries that depended on coppicing, and these can still be identified today. The presence of charcoal-burning hearths or platforms (known as 'pitsteads' in some places) is perhaps one of the best known and most widespread, especially in areas where charcoal was needed as a fuel for smelting iron ore; for example, the Lake District, the Forest of Dean, the Weald and the woods around Sheffield and Telford. These platforms were in use mainly between 1700 and 1900, although for most, their life was much shorter, and some were probably only used for making charcoal on a handful of occasions. However, the charcoal was mainly made from coppiced material and so their presence is a good indicator that the wood in which they are located was coppiced in the eighteenth and nineteenth centuries.

*Very old hazel coppice wood in desperate need of restoration.*

*Dense sweet chestnut coppice.*

### *Species' Composition*

Species' composition may or may not be important – so much depends on your objectives. If your sole reason for coppicing is to produce firewood, the optimum make-up of species will be completely different to that if your objective is to produce material for chair-making or hazel products. For example, a sycamore coppice just won't be suitable if you want to make a wide range of craft produce to take to local shows. Conversely, a pure hazel coppice in rotation won't supply the needs of even a domestic firewood demand, even though it would be a joy to a hurdle-maker. For sheer enjoyment, variety of habitat and wildlife, a mixed coppice will provide a good variety of raw materials for most coppice workers.

*Plantation on an Ancient Woodland Site (PAWS); Norway spruce on the right with some remaining coppice stools to the left wood.*

### *Softwoods versus Hardwoods*

You may find you have little choice but to take on a wood that has had some softwood planted within it. Apart from the fact that softwood will not coppice, this may not be too much of a problem if a ready source of low-grade firewood is required. The wildlife benefits of converting a Plantation on an Ancient Woodland Site (PAWS) back to mainly broad-leaved species are overwhelming. The aesthetic considerations, biodiversity benefits and the ultimate increase in value of the timber in your wood are just some of the reasons to steadily thin and eventually remove most of the conifers. However, even in ancient woodland, a few scattered larch, Scots pine or even something more exotic is not normally a problem, as long as it poses no threat to the native species by spreading at their expense.

### *Native versus Non-Native*

No book on trees is complete without at least a paragraph on native versus non-native trees. There are plenty of books detailing which species are generally accepted as being native and those which aren't. Some species have been given the term 'honorary native' – those that have long been naturalized; for example, sweet chestnut and sycamore. Conservationists have heated debates over whether a particular species is native or not. It is quite possible to argue about why a species seems to be quite at home in one part of the country and not another, or indeed lives in one wood but not the next wood. Beech has long been thought not to be native in Northern England, but there is increasing historical evidence that it might be. Several popular species have been heavily discriminated against because they have not been regarded as being native in some places, including beech, Scots pine, sycamore and sweet chestnut.

To some extent, many of these arguments are becoming increasingly redundant now as climate change exerts its influence. The talk now is of 'frontier species', i.e. those that are moving north as a result of global warming. Perhaps the best example is holme or evergreen oak (*Quercus ilex*), which, having been introduced from southern Europe in the nineteenth century, is now spreading along the south coast of England under its own steam.

We do need to be thinking very carefully about the influence of climate change on our woodlands. A tree planted now won't be mature for at least forty years; by that time, the climate may well be completely different to what it is

*Eucalyptus coppice – high yielding but of little benefit to wildlife.*
*(Photo: Andrew Leslie)*

*Mixed coppice after one year's growth.*

now. We need to consider whether we should be establishing woodlands composed of species that are more tolerant to summer drought and perhaps take the opportunity presented by less frequent severe winters; for example, some of the eucalypt species, many of which coppice very well.

For perhaps twenty years, those planting trees have been encouraged to use native species from their region, according to a system devised by the Forestry Commission. Speaking from a North-west England perspective, it has often been difficult to obtain Cumbrian birch or Lancastrian oak, but over the past decade, this situation has improved. The point of this was to ensure locally sourced genetics, assuming that biodiversity dependant on natural Cumbrian birch, for example, would readily use new habitat provided by planted Cumbrian birch, rather than planted birch from say, Suffolk or the Czech Republic. However, the experts on climate change advise us that we should be making our woods more robust and adaptable to this change. What does this mean in reality? Well, it could mean going against the advice of the past twenty years and planting trees from a wider variety of provenances, and perhaps even going to the extreme of avoiding local provenance. In this way, we may be able to provide a wider gene pool for native trees and shrubs to generate new phenotypes, which could be more tolerant of or better adapted to the predicted warmer and drier summers, and milder and wetter winters of the future.

## Appropriate Management

It is important to consider the implications of coppicing on the wildlife of the wood. Generally, if the wood has been coppiced in its recent past (the last fifty years), the chances are it will have a long history of coppicing. Many (perhaps most) Ancient Semi-Natural Woodlands (ASNWs) have a long history of coppicing. This means that on balance, the wildlife in such a wood may be at least partially adapted to coppicing and its cycle of light and shade. Any research you can do to identify important species that may be in the wood will help your future management decisions. Are there signs of dormice? Are there any unusual lichens that may be adversely affected by coppicing? You may have to rope in a specialist ecologist if you are anxious to take into consideration the needs of rare and

*Fallow doe. (Photo: Teresa Morris)*

endangered species within your wood. However, if you do engage a specialist to search for interesting or rare species and one or more of these is discovered in your wood, you then have to consider whether you should, and how you could, alter your proposed management. If your woodland has an uncommon lichen, for example, but it has clearly been coppiced in the past, it is worth remembering that that lichen has survived such management in the past. Most woods, if they are big enough, can accommodate a balance of management styles, so that you can have a coppiced area, a non-intervention area and an open area, all within the wood. A mix of structure and approach should maximize your wood's biodiversity.

(Refer to the section in Chapter 3 about designated sites and European protected species legislation.)

### Deer

The presence of deer in your wood should not affect your decision on whether to purchase, lease or manage the wood. Of course, it's wonderful to see our largest wild animals in the countryside, but their presence will affect all your woodland management decisions, which could well cost more than if deer were not present. If your woodland is just one of many in the area, what you do regarding deer management in your wood alone probably won't make much difference to the general way deer behave in the area and the damage they may or may not do. One thing is certain though – if your wood has deer in it and you are going to coppice your woodland, you need to consider all your options very carefully. Please see Chapter 4 for more on deer.

*Coppicing can be great fun for families.*

### *Further Research*

There will probably be a limited amount of research you can do before you purchase, lease or work your prospective wood. But if you do have some time and you are serious about the project, it might be possible to find ecological and/or historical information about the wood from a number of different sources, and this information might help you to decide on a management strategy. For example, local record offices may hold information on the sale of timber or underwood, or of the wood itself; the deeds could have some useful information in them, but it's unlikely you'll be able to see these before you purchase. Natural England, or in some cases, National Parks or Area of Outstanding Natural Beauty offices may hold information on the ecology of the wood (sometimes known as Phase One or Phase Two surveys). The County Wildlife Trust may hold information if the wood is a County Wildlife Site (normally the county trust administers these, but their description varies according to the county you are in). If the wood is a Site of Special Scientific Interest (SSSI), there will be a citation – the reasoning behind its designation – these are publicly available for all SSSIs on Natural England's website. You may be able to find out more about woodland's history by looking at old maps and talking to local people, though often past coppice management has faded into a distant memory for all but the oldest members of a community.

## TAKING THE PLUNGE

Having looked into every nook and cranny of your prospective woodland and weighed up the inevitable pros and cons and decided that it is the right one for you, then now is the time to take the plunge. Whether it involves lengthy negotiations with the vendor or lessor, or a tense battle of wills at an auction, you will now have to put your money where your mouth is and go for it!

In the next chapter we will look at the legal implications of your acquisition and make suggestions for how to get started with your new life as custodian of the woods.

# Chapter 3
# What will I Need?

## INTRODUCTION

You may be keen to get started at coppicing but before you do, take the time to consider these important questions: What is the best way to manage my wood? What tools and equipment will I need? What do I need to know about my legal obligations, and what optional things should I do to enhance my experience of owning a wood?

## LIVING WITH YOUR WOOD IN THE EARLY DAYS

Once you have secured your woodland to work in, don't rush in to do lots of work straight away. Much depends on the time of year when you sign the lease or exchange contracts, but a really good piece of advice is to live with your wood for a while. The length of time you do nothing should ideally include a spring season. There are many reasons why you should show restraint! Firstly, you may need to obtain permissions to carry out tree felling, coppicing or other work; you might need to develop a working relationship with a neighbour; you may wish to work with a woodland adviser, ecologist or other expert to get the right advice. Perhaps the main reason to delay starting work is to make sure that work is properly planned and considered – you may regret acting in haste. Problems that could arise include damage to archaeological features, biodiversity, the view – both into and from the wood – or access. It's a cliché but once a tree is felled it can't be put back up again (although coppicing is about the closest you can get to this). One of the authors remembers being told off by an eminent botanist for stacking firewood on a trackside; the area covered in wood was the best stand of an uncommon wildflower in the area, and so a great deal of effort was put into moving several tonnes of wood to avoid further damage to the plant.

## MANAGEMENT PLANS

A good way of steering clear of making bad woodland-management decisions is to write a management plan. Having a plan can help you work out the best options for a wide range of activities in the wood. Sometimes, the documenting of people's thought processes are as important as the eventual decisions made. Just carefully thinking through the options and weighing up 'pros and cons' will help you decide on the best route to managing your wood. The management plan can be whatever you want it to be: it could be a lengthy, complex document, or just a page or two of options and bullet points. It could even be a simple annotated map. Your plan should suit you, and be something that you refer to from time to time – there is little point in writing a plan that gathers dust on a shelf. One of the benefits of a plan is that if you sell the wood or leave it to a relative, the plan will help them understand why you chose a particular management option, and it will help to inform future management decisions in the wood. In years to come, it will also remind you why you yourself took that particular management route.

There have been many management-plan templates outlined in the past but the main points you need to cover in a plan are:

- Description of location, including designations.

- Description of surrounding land uses and links to other woodland.
- Description of physical features such as soils, altitude, aspect, slope, climate.
- Description of the important features of the wood – this could include flora and fauna, specific trees such as veterans, pollards, rare species, archaeological features, views and so on.
- Barriers and constraints to management – these could include designations, rare or protected species, steep slopes, rocky or wet areas, invasive or unwanted species, access, boundaries and rights of way.
- Woodland management history.
- Your objectives.
- Woodland-management options.
- Costs versus resources (including a cash flow forecast, if you wish).
- Map(s).
- Appendices could include historical maps and information, old photographs, interview notes with former owners or managers, records of your own recent management activities, copy of felling licences, grant-aid contracts and other agreements and contracts.

The Forestry Commission has a management-plan template that is freely available to anyone; this is normally used by those who have a woodland grant scheme management-planning grant, and it is designed to cover the areas required for woodland certification purposes. It won't suit everyone's idea of a plan but there are some useful ideas in it. If your woodland is used by the public, or is particularly close to a community, it is a good idea to involve them in its management. As a minimum, this can include explaining what you are planning to do, the rationale behind this and when the work is going to take place.

## EQUIPMENT

Take your time to decide what equipment is essential. The right combination of hand-tools and mechanical help will be as varied as the woods that you are managing and will be governed by your inclinations and your purse, so do not rush in to purchase 'toys' that will later prove obsolete.

*Managing a wood need not be all chainsaws and machinery – here, cooking twist dough bread over a woodland camp-fire.*

Many people like the idea of working in the woods and enjoying the tranquillity that is brought by the birdsong and the noise of the wind in the trees. There is something quite primitive and enjoyable about the sound of a tree being felled by hand. However, a coppice business would struggle to survive in the modern day without the efficiency of felling by chainsaw, and there will still be many occasions when you can sit down and enjoy being in the wood without the noise of the chainsaw.

*Chainsaw and associated equipment.*

Buying a chainsaw will probably be your main essential purchase. This is an expensive business and so the buying decisions you make should be given detailed thought. You may be tempted to pick up a cheap second-hand saw, but remember that a chainsaw is a very dangerous tool. It is essential that all the safety features are modern and working properly, and it is most important that you also buy all the recommended protective clothing. This should fit properly and be comfortable – there's nothing worse than trying to concentrate on felling a tree when your boots are giving you blisters. Your chainsaw should be appropriate for the size of tree you'll be felling; don't be tempted to get something too big and heavy. There are a couple of well-known brands of chainsaw that, although more expensive than some of the lesser known makes, are heavy duty and designed for daily use; the smaller, lighter chainsaws tend to be cheaper but are not as robust.

Chainsaws are oily, messy machines and so you'll need somewhere secure in a workshop or garage where you can keep the saw, fuel can, spare chain oil, two-stroke oil and toolkit. If you intend to start a coppice business, there may be other powered tools or equipment that you will need in due course. If you know you have an identified firewood market, then a firewood processor might well be an early investment; following on from this, you'll probably need a vehicle to deliver firewood in or to tow a trailer. There are a few examples of coppice businesses in unusual situations where customers come and collect their firewood, thus negating the need for a delivery vehicle. Other equipment will depend very much in what you decide to specialize; some businesses have a chainsaw mill that can be used to plank up standards or other large trees felled in the wood.

*Chainsaw mill – useful for adding value to small standards.*

**Financial outlay for basic chainsaw equipment and training**

| *Equipment* | *Approximate cost* |
|---|---|
| Small chainsaw | £200 |
| Toolkit: basic (files, grease, spanner, spare chain) | £30 |
| Combination fuel/ oil can | £25 |
| Helmet with visor and ear muffs | £35 |
| Trousers | £85 |
| Gloves | £25 |
| Boots | £60 |
| Felling bar | £45 |
| First-aid kit | £25 |
| Basic chainsaw course (Units CS30 and 31) | £700 |
| First-aid course | £85 |
| **TOTAL** | **£1,315** |

### *Harvesting and Extraction*

Wood is heavy, especially wood that you've just cut, because it's full of water. Therefore, unless you are using the wood where it's felled, you will need a method of getting it from A to B. There are many possibilities here and, potentially, you can spend a great deal of money. Even if your woodland is within a stroll of your back door, you will need a means of getting the wood to where it will be used. Options range from a wheelbarrow to a four-wheel drive forwarding machine.

You may consider buying, borrowing or hiring equipment, or alternatively paying someone who already has the right equipment. In our experience, this is a complex choice. You certainly need to think long and hard about buying an expensive piece of equipment that could easily become an unused white elephant.

Borrowing machinery sounds fine, but what happens if it's in an unsafe condition, or if you damage it? There are likely to be insurance implications. It's bound not to be available when you need it. The same goes for hiring machinery; of course, this is less expensive than investing in your own machinery, but the condition of hired equipment is often poor; never hire a chainsaw.

Paying a forestry contractor may well be a good option – and relatively cheap compared to purchase of machinery. You may be able to find one that comes with good references or that works for a nearby estate and who, therefore, can't afford to loose his reputation. A forestry contractor will probably have sizable machinery and you may have to accept some damage to access routes, entrances and trees that line tracks. However, it should all be over in a short period of time, and most harvesting damage will green over quickly. If you're really worried about damage to the wood, you can write reparation procedures into the contract – always have a contract!

The other option is to purchase equipment to extract your wood. The romantic idyll of having a horse to 'snig' your wood from the cutting area to the processing area seems wonderful. If your heart is set on this, then it is imperative that you (and your horse) receive the correct training – contact the British Horse Loggers Association for assistance. Purchase and maintenance of a horse is at least as expensive as purchasing and maintaining a piece of inanimate machinery and you need something to transport both equipment and horse. Using a horse, however, is much quieter and has a lower carbon footprint.

Perhaps the most traditional means of extraction is a small tractor – old or modern. You can buy a serviceable vintage tractor such as a Fordson Major for a little under £1,000. These have been used for decades in forestry and are reliable workhorses, but in a coppice situation, they are cumbersome, unwieldy and very heavy. Perhaps the best way of extracting wood in these modern times is an All Terrain Vehicle (ATV or quad bike). Even a second-hand ATV can set you back well over £1,000, and a brand new one, over £7,000. Ground clearance is not perhaps quite as good as a larger tractor, but if you cut

*Harvesting and extraction options: (a) tractor and timber crane;*

*(b) pickup;*

*(c) extraction with horses.* (Photo: Ted Wilson)

*ATV or quad bike – perhaps the most flexible option for harvesting wood.*

your coppice stools nice and low, this shouldn't be a problem. If there are any slopes at all in your wood, a four-wheel drive ATV of an engine capacity of at least 350cc is recommended above two-wheel drive.

Clearly, if you live any distance from your wood, you will need a trailer to transport the quad bike in; if you are working out a budget, don't forget that you will need a tow bar fitted to your vehicle.

It is possible to strap wood to your ATV but on slopes this can be dangerous, and in any case, this is rather limiting. A better option is to get a simple sheep trailer to go behind the ATV – this is much more versatile and quite long lengths can be carried in this way; perhaps the only limitation is the weight of the piece of wood you can lift on to the trailer. There are several implements that can be purchased to tow behind your ATV, including one to allow you to extract larger lengths of timber or a trailer with a timber grab.

There are many coppice businesses that regard a four-wheel drive pickup as being the ultimate multi-use vehicle. Pickups are versatile and flexible, providing both a means of extracting produce from the woodland, and a means of delivery. Similarly to the ATV/trailer set-up, the only limiting factor is the weight of the piece of wood you can lift. There are small cranes that can be fitted behind the pick-up cab to assist with loading and unloading heavy produce.

### Tools

This section deals with hand-tools – you don't always need a chainsaw in the woods! In the same way that felling and harvesting equipment can be expensive and potentially a big drain on resources, it is also possible to spend a great deal of money on hand-tools. If your main product or business is likely to be firewood or charcoal then you probably need no hand-tools at all. However, if you have a creative streak and plan to make use of much smaller dimension wood, then there is the potential to amass a huge variety of interesting and useful tools. Some tools have been used by woodworkers for centuries and it's quite likely that a Roman carpenter

*Gränsfors Bruks side-axe.*

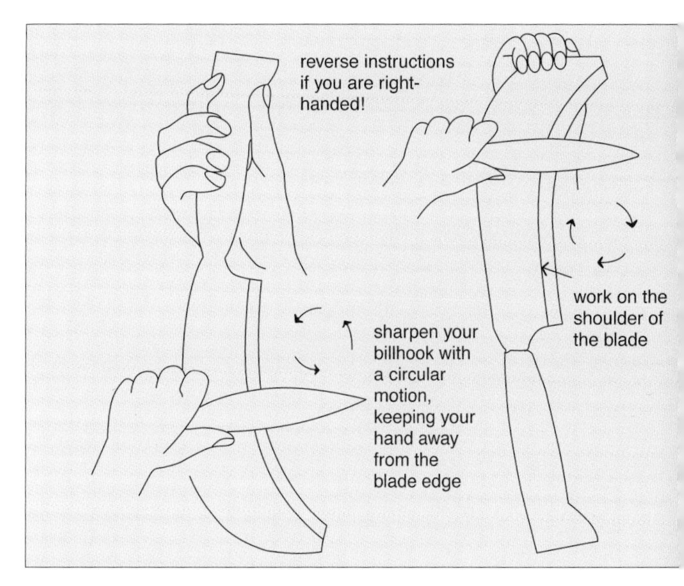


*How to sharpen a billhook.*

would feel quite at home in a modern woodworker's workshop. It is possible to fabricate some tools yourself, but this is a skill that is beyond the time and knowledge resources of most people.

The chapters on coppice crafts will cover some of the more specialist tools. For basic woodland work, there are three principal tools you need: a billhook is probably the most useful, followed closely by a bowsaw and a sharp knife. Loppers are an invaluable modern addition, and these are especially useful for those not yet confident with the use of a billhook.

The billhook is an ancient implement used throughout Europe; the 'bill' is a blade, and the suffix 'hook' was added as many designs include a hook at the blade tip. There are many, many different designs that have developed over the centuries, mostly on a regional basis; different types were used for slightly different uses – for example, some hooks are known as hedging or thatching hooks. Famous makers include Elwell and Nash, who produced hooks that are still in daily use today and are sought after; generally, the steel was of higher quality than the more modern tools, and could be sharpened more easily into a better edge. There are modern, new billhooks available for about £25, and these are better quality than those of recent years, which were often cheap imports.

A billhook is the essential tool for dressing out hazel. In many regions it was the tool used for harvesting hazel rods – they were cut from the stool with an upwards motion. This tends to leave a rather pointed stub on the stool and these days, a flat cut is preferred. In order to keep your billhook sharp and therefore efficient and safe, you will also need a sharpening stone.

Billhooks are frequently used to split hazel for hurdle or spar making; these are usually the smaller hooks. The large and heavy Yorkshire hook has a blade on both sides; the flat blade was thought to be used for trimming besom broom heads.

A billhook can be used to reduce the length of rods but this can only be done efficiently up to a point. Once the rod reaches a certain size, a bowsaw (sometimes known as a 'bushman') is the more appropriate tool. These are the tool of choice for coppicing small stools by hand. The small bowsaws have a pointed nose making it easier to get the saw amongst the coppice stems. Bowsaws are widely available; remember to keep some spare blades handy.

A saw well worth considering is the modern pruning saw. These normally cut on the back-pull of the sawing motion only; they are ferociously sharp when new and the blades last much longer than a bowsaw; the blade sometimes folds away when not in use, so these can be carried safely in

**The more commonly used tools and their approximate cost**

| Tool name | Use | Approximate cost |
|---|---|---|
| Drawknife | Reducing cleft wood to rough rounds prior to turning; preparing other materials such as removing bark or shaping. | £35–£80 |
| Spoon carving knife | Carving spoons | £15–£35 |
| Axe – hatchet | Reducing wood to smaller dimensions. | £15–£50 |
| Side and carving axes | Reducing wood to smaller dimension and carving (not tree felling). | £50–£80 |
| Adze | Making chair bottoms, carving out large bowls and giving oak beams a particular traditional medieval finish. | £70–£150 |
| Drill and augers | Making holes in wood, mainly for joints. | Drill £45; augers £10–£150 |
| Froe | Used for clefting wood and timber. | £30–£80 |
| Shave horse | Simple wooden, sit-on vice, used to grip wood when being worked on with a drawknife. | £220 – or make your own |
| Rounder/rotary plane/ stail engine | Rounds sticks to make handles (stails) for rakes, brooms, etc. and chair components. | £30 |
| Spoke-shave | Helps to accurately shave wood to specific shapes, including chair bottoms. | £45 |
| Loppers | An easy way of reducing rough ends or chopping to length. | £25 |
| Twybill/morticing knife | Removes wood from mortices. | £60 |

a pocket. These were developed for foresters who would prune the lower branches from living trees to promote better timber quality and to allow access through dense forest.

You may opt to invest in a larger cross-cut saw – these can be either one-person (a handle at one end only) or two-person (a handle at both ends) saws. These can fell trees remarkably efficiently when sharp, but it is tiring work and inappropriate for the modern coppice business. Sharpening a cross-cut saw properly takes some time and a lot of practice.

A sharp knife is such a useful thing to have in the wood, that we consider it indispensible. Most coppice workers will be bundling up produce from time to time and having a knife handy for cutting twine is essential. One of the most popular knives is the wooden-handled French 'Opinel' make, available in a range of sizes – these fold away like a like a traditional penknife. These aren't designed as carving or whittling tools, which have a shorter, fixed blade kept in a sheath – most people use 'Frost' carving knives for whittling.

There is a variety of kinds of lopper – some have a scissors action and some an anvil action; some have extendable handles so you can reach higher and some have a ratcheted action that enables you to cut thicker stems. Loppers are a handy item for making neat, flat cuts and for reducing rods to a length with a particular degree of accuracy, and for trimming the ends off in certain crafts (see Chapter 6).

So, those are the most popular tools considered essential to those working in the woods but there are many more kinds of tool, some of which are useful for a range of activities and others which are used less frequently for much more specialist tasks.

There isn't the space to discuss the more specialist tools here, although some of them are mentioned in subsequent chapters. They include bark-peeling irons, travishers, stock knives, all the tools associated with the pole-lathe, including hundreds of different chisels, wood carving chisels and so on.

Many tools have been left in the woods over the years. There really is no easy answer to this, except to check thoroughly when you leave the wood that you have all your tools with you. You

*The most useful coppice worker's tools: (a) drawknives;*

*(b) froe (above); (c) billhook (below).*

*Array of coppice craft tools.*

can paint the handle of tools with some bright paint or wrap them in fluorescent tape and this does help to locate them amongst the leaves on the woodland floor. Tools should not be hung up in trees, as tempting as this is; they can easily be dislodged and fall on you. Equally, a billhook should not be left sticking up out of a stump – this damages the stump unnecessarily and you can easily trip over it or fall onto it. Many people use a leather holster in which to place tools and this is probably the safest option, and that way, you always have your billhook to hand. If you have a few tools in use lay out a tarpaulin and make sure that the tools are placed on it when not in use.

## Other Useful Items

There is a range of other things that you'll find useful at some point in your coppicing career. Some of this stuff can be kept in a tool store or shed, but most can be transported to site whenever you need them: things associated with eating, such as a kettle, basic utensils such as a plate, cup and fork; you already have your knife with you; you might want a water carrier to take water to the wood, or collect water in the wood.

There are items you will need that are associated with keeping warm and cooking, including matches (or a fire-steel if you're being ultra-green); kindling is best taken with you from home; if you visit your wood regularly, you'll soon run out of easily gathered natural kindling material. Although those with some bushcraft knowledge can easily fashion cooking stands and so on from coppice materials, it will save time if you have the basics to hand, such as a stand to put the kettle on or a tripod and hook to hang it from. Remember to keep some dry logs or brown ends from a charcoal burn in the wood, so you have some firewood that catches quickly and burns well.

If you're not planning a fire on which to heat water, and you're not keen on using a flask, a storm kettle is a useful addition to your equipment inventory. The most common make is the 'Kelly' kettle, which comes in a range of sizes. A small fire is made and the storm kettle placed over it, so the heat and smoke exit from the top; the water is contained in the sleeve around the kettle and boils extremely quickly; the fire can be made from a small quantity of dry twigs and leaves and additional fuel can be added through the smoke hole in the top.

A chair is usually more comfortable to sit on than the ground, especially if the ground is frozen, and it's a bit easier if you're over forty! Old garden chairs are easily stacked away in a shed or under a tarpaulin.

*Kelly kettle.* (Photo:Warren Sanders)

### *Shelter*

Permanent shelter within your wood is not essential. However, many people create some kind of construction; some of these are temporary and some end up being more permanent (see later in this chapter for more on planning restrictions). Garden sheds are functional but often relatively expensive and sometimes look horribly out of place in a woodland setting, especially when new. An old shipping container could be used to store tools and materials – they are weather-tight and secure but, again, they look very ugly anywhere in the countryside. Some people prefer a more traditional shelter, such as a yurt or tepee. These can be expensive to buy, but you can make your own.

The simplest shelter is one made with a tarpaulin and some rope. These can be put up and taken down quickly on the day you're in the wood; the 'tarps' can be cheap to buy and it doesn't matter too much if they get battered or ripped after a few uses. A basic tarp shelter can be put up using surrounding trees and it keeps off the worst of the wind and rain from a work area. They can be made to last more than a day or two by erecting a simple wooden frame from poles cut in the wood. However, you do need to be careful that the local planners don't think you're starting a more permanent building; make sure the shelter can be defended as being temporary.

*A simple tarpaulin shelter can make a very effective workshop area.*

*Coppice course in the woods.*

Even if you plan to work on products away from the wood, you will probably still need a working area or 'yard'. The location for this should be chosen carefully. Ideally, it should be adjacent to an access track to enable ease of loading and unloading; it will probably be not too far from the entrance to the wood from the road, as you may want to access your yard in all weathers. It is probably best if the area is out of sight of the road to avoid undue attention from unscrupulous passers-by. Having a flat working area will be important in order to stack wood safely, carry out charcoal burns, erect shelters and, again, for ease of loading vehicles. In addition to all this, the yard should not be placed in the best wild-flower area in the wood. You might only make a little mess in the first year, but even occasional weekend working will gradually destroy the ground flora, and anything more than that will potentially generate a morass of mud. Of course, the way around this is to lay down a stone surface, and although this can be a good solution if carried out correctly, it can make problems if done badly – some areas could be made wetter by putting stone down. If your wood is ancient, you should choose a stone of the same acidity as the existing conditions; otherwise, the ground flora surrounding the stone area will gradually change over time to reflect the change in acidity.

An alternative is to use wood chips, which are generally more environmentally friendly but unless the layer is deep, this won't help with very wet conditions. There is a train of thought that says that a new track or working area established in an ancient woodland, as we've described here, amounts to a loss of that ancient woodland, and so this kind of up-grading of access should be considered very carefully.

### *Toilets*

And now to the delicate subject of woodland loos. Clearly, if you are a very occasional visitor to your wood, you won't need to think about this. However, if you visit every week or every day, or if you begin to run regular events or courses, the issue of toilets could become important. A hole in the ground is probably the simplest, and a makeshift cubicle can be rigged up quite easily; the soil from the hole should be close to hand, with a small spade – this is used to cover up what's been put in the hole. This is rather primitive, however, and may not suit some people if you start running courses. An easy alternative is to use a portable toilet from a caravan, which can be taken away and emptied where appropriate. There are options that are more long-term but planning permission would be required for a permanent toilet. Just as with work areas and tracks, if every ancient woodland owner installed a compost toilet in their wood, the nation would lose dozens of hectares of irreplaceable ancient woodland habitat, so try to get a balance between encouraging use of woodlands for courses, events and forest schools, and the associated development and damage that sometimes goes with these activities.

## TRAINING, HEALTH AND SAFETY

### *Courses*

People tend to be very happy to spend their money going on courses they know they will enjoy. These are usually the craft courses and others related to basic woodland management. Courses people tend to neglect are those that could help them to do work more safely, efficiently or that could even save their life! There are many useful courses available; some of which may come in useful, including hand-tool maintenance, tree identification, basic woodland ecology and woodland management.

Please don't ever use a chainsaw without having attended a course that covers basic cross-cutting and maintenance as a minimum (Unit CS30); if you are felling trees other than very small coppice, you should also attend the felling small trees course (Unit 31). Chainsaw technology and techniques in their use continue to develop, and even if you have had some training in the past, it's worth keeping up to date. A basic course will also tell you whether the chainsaw you picked up cheap at that farm sale is legal and safe or not. A second plea – please don't ever use your chainsaw without wearing the protective clothing – even to cross-cut firewood. The protective trousers, boots and helmet with ear muffs are essential at all times when the saw is running. Don't be tempted to use hand-me-down protective clothing. The design of protection in chainsaw trousers has improved dramatically over the years, so throw away those old protective leggings and spats and buy something new. Some people think that a garden or electric chainsaw is somehow safer than a professional model used in the forest. Obviously, this isn't the case; if it will saw through a six-inch log, clearly it will saw through your leg. Never hire a chainsaw, as you don't know its condition. Always keep the cover on the saw, as even when stationary, the chain can give a nasty cut.

> The only safe chainsaw is one with no petrol in it, no spark plug, and on a low shelf so it can't fall on you.
>
> Seale-Hayne College (1982)

Without wishing to sound macabre, perhaps one of the most important/essential courses you should single out is first-aid. Chainsaw injuries can be extreme and simple knowledge of how to stop flow of blood can save someone's life... or your own. There are plenty of other common injuries that will need simple first-aid in the wood, including cuts from saws and billhooks. There is also the unexpected, such as a branch falling onto a head (even if the head has a helmet on it!) and things like twisted ankles (all witnessed happening in various woodlands by the authors).

### *All Terrain Vehicles (ATVs)*

ATVs are potentially dangerous vehicles – people die every year through careless driving and making simple mistakes. A course on driving your ATV is recommended; always wear a helmet. Some ATVs are supplied with a short safety film – do watch this – there are some salient

*Encourage participants on your work parties to wear protective clothing.*

stories of experienced ATV drivers having serious accidents, some of them resulting in death. The most common mistake is to underestimate either speed or slope and sometimes both. This results in the ATV rolling over – with no roll bar on an ATV, serious consequences are likely, such as the machine rolling on to you.

If you are using any other kind of machinery, there is usually a course on its use, though sometimes courses can be difficult to find. It's a good idea to contact your local agricultural college, which will often have a programme of courses that farmers may need, including ATVs, tractor driving and towing a trailer.

## Public Footpaths

If you are working close to a highway or public footpath, you need to erect the necessary signs warning people that tree-felling operations are taking place. Most machinery stockists or protective clothing companies stock the kind of signs you may need. If you have to fell across a public footpath, it's best to have two helpers to keep a lookout for walkers whilst the actual felling is taking place.

## Work Parties

More health and safety. This is not the most exciting subject but it is important for your own peace of mind, and possibly to meet insurance requirements, that you consider how to avoid accidents and to minimize risks. A basic risk assessment isn't difficult or time-consuming to complete; you can use the risk-assessment table in Appendix II. Know how to describe your location, including its grid reference, and where the nearest hospital is. If you have no mobile phone reception in your wood, know where the nearest public phone box is, or friendly farmer – have a plan that you can implement easily if there is a serious accident.

You should ideally have a first-aid kit to hand; preferably also have someone present who has attended a first-aid course recently. Have a supply of clean water handy to wash cuts and wounds. If a chainsaw is being used, the minimum requirement is a large wound dressing to staunch blood flow.

At the start of the day, give a 'tool talk'; take people through basic techniques and show them how to avoid the simple mistakes like hitting their head with a billhook. Hand out safety equipment; you can't make people wear it but if they don't, that is then their choice. This could include heavy duty gloves, hard hats and steel-toe-capped boots. Make sure your tools are well maintained – handles properly attached to blades; blades nice and sharp.

If there is a chainsaw user, then that person should be completely in charge of their work area. They should check all around to ensure people are a safe distance away – usually recommended to be two tree lengths. Don't light your fire or have your base camp within the reach of the chainsaw user's falling trees, as that really is tempting a disaster.

If there are deer in your wood, there are likely to be ticks (*Ixodes ricinus*) on the vegetation. The ticks can spread Lyme disease which, if untreated, can ultimately result in nasty symptoms, although the disease is uncommon. After each visit to the wood, you should check yourself for ticks. Warm, moist areas such as the groin and armpits are the areas where ticks tend to migrate. If you find a tick, it should preferably be removed with tick tweezers; if this is done within 24h of becoming attached, the risk of infection is greatly reduced. The commonest early sign of an infection is a halo-shaped rash spreading from the location of the bite. If you think you may have found such a rash a few days after a woodland visit, especially after knowingly having a tick attach itself, you should visit your GP. Antibiotics are effective if the disease is caught early, but you may have to tell your GP about Lyme disease as it is not common. The advice for prevention is to wear close-fitting clothing and to tuck socks in boots; insect repellent may also deter ticks from biting. However, the main way to prevent infection is rigorous checking for tick bites.

## WHAT YOU CAN AND CAN'T DO IN YOUR WOOD

### *Designations*

Some of this may seem incredibly obvious but sometimes the most obvious things need to be stated. There is quite a lot of legal stuff that needs to be covered here. If you think '…this doesn't include me' or 'I don't need to read this', then think again please. No matter what your objectives, you need to stick within the legal framework set out for woodlands and forests, and you don't want to set out on the wrong note with the various authorities.

There are number of wildlife designations, although none are specific to woodlands. Some sites are designated according to European legislation and some according to UK legislation. It is quite unlikely that you would end up looking after a site with a very high-ranking designation, such as a National Nature Reserve (NNR) or a Special Area for Conservation (SAC), for example, so don't let these designations worry you. It is possible that your site could be a Site of Special Scientific Interest (SSSI). These are often privately owned and represent the best examples of their habitat type in the area; they may be of national, regional or of county-wide significance. Regardless of whether they are of national or more local significance, these sites have been described as being the crown jewels of habitat in Britain.

If you're managing a wooded SSSI, you will need to discuss all management operations with Natural England, which designates and administers them. This means you need to discuss any operation including tree planting, felling (including coppicing), access work and so on. However, it is still relatively unlikely that your woodland will be designated an SSSI.

Most local authorities in the UK operate a further layer of designations; these may be run by the local authority or the county wildlife trust, and have different names – here in Cumbria, they are called County Wildlife Sites. These sites have no statutory authority managing them or restricting what you can do, but represent a further tier of habitat 'ranking'.

As described in Chapter 5, the term Ancient Semi-Natural Woodland (ASNW), has become

*Woodland in a protected landscape – the Lake District National Park.*

much more important over recent years, to the extent that in many areas, the authorities regard this term as a designation. This is despite the fact that the Inventory of Ancient Semi-Natural Woodland, maintained by Natural England, continues to be called 'Provisional'. Countryside and woodland pressure groups have been largely successful in getting this designation recognized by the planning system. In terms of development in the countryside, this is vitally important, and in terms of how you manage your woodland, it may greatly influence your decision-making.

SAC
NNR
SSSI
County Wildlife Site
Ancient Semi-natural Woodland
Woodland in a designated landscape
Older plantation woodland
New plantation woodland

*The top of this triangle represents the small number of sites with high ranking designations and the broad bottom represents all others. SAC – Special Area for Conservation; NNR – National Nature Reserve; SSSI – Site of Special Scientific Interest.*

### European Protected Species

Certain species of animal are protected by European law. The regulations of 1994 were amended in 2007 to strengthen this protection. This legislation removes the defence of accidental damage to the species or its habitat. Unfortunately, the administrative regimes are different in England, Scotland, Wales and Northern Ireland, and this has added to the immense complexity of this legislation. It is not possible to cover this in any detail in this book, but it is important that you familiarize yourself with the basic principles. The species covered, which are most likely to be in your woodland, are all the bat species plus dormouse and otter.

It is an offence if you damage or destroy a breeding site or resting place of a protected species (even if this is unintentional or the animal isn't present); it is also an offence to kill or injure a protected species or disturb it in a manner that significantly affects it's ability to survive and breed, or, as a consequence, significantly affects the local population.

The Forestry Commission and Natural England (in England at least) have taken the position that they will produce guidance and that prosecutions are unlikely. However, you should follow that guidance if you suspect that any of the species are present. The box below will help you through the compliance process.

It is important to state that this is really only a simplified version of the guidance and if there is any uncertainty, Natural England (or the equivalent bodies found in Scotland, Wales and Northern Ireland) should be consulted. There is detailed guidance available from them on how to proceed with management operations in order to take into account the species involved.

### Landscape Designations

There are several ways in which your wood may be affected by landscape designations and they could affect how you go about managing your woodland. If you're lucky enough to end up looking after woodland that lies within a National Park (NP) or an Area of Outstanding Natural Beauty (AONB), the planning restrictions are likely to be tighter. However, there are often more resources available (special grants or enhanced rates on existing grants) in these areas, so there may be some benefits too.

#### Safeguarding European protected species

Are there any European Protected Species likely to be found in this location and this type of woodland?
**If yes:** Are they known or likely to be present in this particular wood?
**If yes:** Are the proposed operations or activities likely to involve injury to the animal, cause significant disturbance or destroy a breeding site or resting place of one of the protected species?
**If yes:** Will any of the above be deliberate or intentional?
**If yes:** Can the operations be modified to avoid committing an offence by following good practice guidance?
**If yes:** You can proceed.
**If no:** You will require a licence to proceed with operations.

An EPS licence application must satisfy these three tests:
1. The purpose of the operation is to enhance biodiversity.
2. There is no satisfactory alternative.
3. The operation will not adversely affect the conservation status of the species concerned.

## *Living Options*

There have been many battles over recent years between planning authorities and those who wish to live in their woodland. There are national planning policy guidelines to which local planning authorities have to conform in the development of their own policies. The guidance for forestry is similar to that of agriculture, with a few subtle differences. In most cases, the nub of the matter is whether there is a necessity to live in the woods as part of the livelihood reliant on the woodland in question. The most successful reason used to obtain planning permission to live in woodlands has been the necessity to be with a charcoal burn 24h a day. However, this still only amounts to a handful of cases. Some of these have cost many thousands of pounds and have taken years of tenacity, form-filling, representation in court and negotiation.

Erection of sheds and workshops in a wood falls within 'permitted development', as long as they are 'reasonably necessary' – for example, if the shed is used as a bad-weather shelter, store or office; use as a classroom is not reasonably necessary and is, therefore, not permitted under this legislation. Of course, this phrase is rather vague and in order to determine whether a shed is reasonably necessary, you must give twenty-eight days' notice to your local planning authority of what you intend to do – Prior Notification Procedure (PNP), or sometimes called a Notice of Intent. They can then make a decision on whether what you have told them falls within permitted development, or whether you require planning permission. If your shed looks like a log cabin or something else that you can sleep in, your proposal is likely to receive a refusal. A caravan used for forestry purposes is exempt from the PNP, as long as it is not for residential use.

It is a myth that if your structure is temporary, it is exempt from these planning rules. If you plan to use a temporary structure for residential use, such as a yurt, tepee, bender, caravan or any other vehicle with or without wheels, planning permission will be needed.

These public battles between woodland businesses and planners have, in some cases, been a terrible waste of resources for both sides, and a pragmatic solution should have been found much sooner. However, in many ways, it is right that planning control over residential development in woods is very tight. Our woodland resource and landscapes could become degraded without planning regulations.

## *Felling Licences*

The Forestry Act of 1967 (as amended) brought in the felling licence regulations, which are intended to control and regulate the indiscriminate felling and replanting of woods and forests. Felling licences are controlled by the Forestry Commission (FC). Some people will be able to coppice without worrying about felling licence regulations, but you could easily infringe them, so here goes!

You are allowed to fell up to 5m$^3$ of timber every calendar quarter, unless the wood or timber is being sold, in which case, the limit falls to 3m$^3$. It doesn't take long to reach this limit, especially if you're working a neglected coppice with large standard trees or lots of maiden birch – perhaps only two or three oak standards or ten birch trees could take you over the 3m$^3$ limit.

There are several exemptions:

- Trees in gardens, orchards, church yards or public open spaces.
- Dead, dying or dangerous trees.
- Trees felled as part of a planning permission or for maintenance of a utilities wayleave.
- Thinning a wood as part of a normal forestry operation is exempt where the trees to be thinned are below 10-cm diameter. Importantly for coppice enthusiasts, coppice growth under 15-cm in diameter is exempt from the need for a felling licence.

For a really neglected coppice without much hazel, some of the stems will be larger than this and so a licence will be needed. Diameters are known as 'dbh' (diameter at breast height) and are measured, not surprisingly, at breast height or 1.3m.

It is best to be covered by a felling licence if you're unsure whether you need one or not. Your local Forestry Commission Woodland Officer will let you know whether you need one. The form is quite easy to fill in, and the licence should come through in four or five weeks. Licences are

*Only 3m³ (105ft³) of timber may be felled per calendar quarter without a felling licence; a large oak standard like this one could easily contain more than half of this on its own.*

usually conditional, i.e. the granting of a licence is conditional on trees be re-established on the site; of course, if you're coppicing, the trees you fell will be re-established from the coppice regrowth. The other benefit of having a felling licence, even if it's a borderline requirement, is that you can show it to anyone who doesn't approve of what you're doing. It shows that you have consulted the Government's advisers on forestry (the FC) and got official approval for the tree felling you're carrying out. Licences normally last for two years.

## Tree Preservation Orders

Before you do any tree felling at all, you need to check with your local authority – either the unitary, borough, city or district council – whether the wood has been protected with a Tree Preservation Order (TPO). If you've recently purchased your wood, this should have come up in the local authority searches. Usually TPOs were placed on individual trees or small groups, but where a wood was threatened with felling by an unscrupulous owner, sometimes the whole woodland was protected by a TPO; these woods are often on the edge of urban areas or places where there has been a threat of development in the past. It's unlikely that a TPO would be lifted and so, if the wood is covered, you will need to work with your local authority tree officer. There is a chance that your wood lies in a conservation area; these were set up to preserve the buildings within, and the setting of, towns and villages. Again, these are designated and administered by local authorities, and you need to work with them in the same way you would for a TPO designation.

## Rights of Way

If there are public footpaths, bridleways or byways open to all traffic (BOAT), it is probably best to live with these. Technically, it is possible to divert a public right of way or bridleway, but it is a time-consuming and potentially expensive exercise, which can be contentious with local people; there usually has to be a very good reason for a diversion to be approved by the local authority. Sometimes a footpath will get in the way of something such as erecting a deer fence in a sensible place; however, there's not normally much you can do about this. You must not block up a public right of way, as this is a criminal offence.

*Taking your own food to cook in the woodland can be a rewarding experience.*

If you manage the rights of way well, most people won't stray from them. Having a footpath through your wood could be an opportunity, firstly to tell people about what you're doing and, secondly, to generate some revenue from sales of products to users of the path.

## MONEY MATTERS

### *Insurance*

In these days of continual fear of being taken to court, all woodland owners and coppice businesses need to carry insurance. The most useful kind of insurance to have is public liability insurance. This should cover you for anything that could happen to a member of the public in your wood, whether they were invited or not, and whether they were on a footpath or not. However, any organized event, such as a work party or guided walk, will probably need additional special cover. Coppice businesses will almost certainly need a kind of public liability insurance that covers teaching, demonstrations and attendance at shows and events.

Coppice businesses will require employers liability insurance if they employ someone; if the business takes on an apprentice, volunteers or other helpers, it will also need this kind of insurance. If you employ a contractor in your wood, you should ensure they also have public liability insurance cover; you should make sure you see a valid copy of their insurance certificate. You should also satisfy yourself that any person using a chainsaw or other equipment is qualified to do so. This may sound like an approach that is unnecessary – however, a contractor without a relevant chainsaw certificate who has an accident would almost certainly not be covered by either his or her insurance, or yours. The Health and Safety Executive take a dim view of this kind of situation. Of course, if you have an accident yourself on your own land, then that is your own responsibility.

The other kind of insurance that you may need to check is professional indemnity insurance. This is usually carried by professional advisers such as Chartered Foresters. This covers damages to them that arise out of giving poor advice. In comparison to other kinds of insurance, it is rarely needed in practice, but asking to see a certificate of professional indemnity insurance is another way of ascertaining the validity of an adviser.

### *Tax*

Ownership of commercial woodland can have some useful tax advantages. The following information is for guidance only. Tax rates and mechanisms may change, and you should take independent tax advice from an appropriately qualified professional advisor on the full tax implications and any consequences particular to your individual circumstances.

Your woodland is likely to be classified as being commercial if you sell timber or wood products or firewood; this does not have to be profitable but you do need to keep a simple profit and loss account and have clear, written objectives for the wood's management (this could be a woodland grant scheme or felling licence). If you simply take firewood for personal use, then it could be classed as amenity woodland and the tax benefits do not apply.

#### Capital Gains Tax

There is no Capital Gains Tax (CGT) liability on the gain in the value of commercial tree crops. Any gain realized on the disposal of woodland is

*If you are a land-owner, income from your timber is exempt from tax until you start to add value to it.*

split between the gain attributable to trees – on which no CGT is payable – and the gain attributable to the land – on which CGT is levied. Where funds realized on the sale of a business asset are re-invested in a new asset up to twelve months before or two years after the sale, then any chargeable gain from that sale is deferred until such time as the new asset is sold. So gains from business assets can be rolled-over by the purchase of forestry land (but not standing timber), thereby postponing any CGT liability.

### Income Tax

Any income or profit generated from your woodland through sale of timber or wood is exempt from income tax, whether the wood is owned personally or through a company. However, income from shooting, stalking and renting the woodland to someone else is liable to income tax. Forestry grants are not taxable (except for annual income from the Farm Woodland Premium Scheme or its equivalent, paid to help off-set loss of agricultural income when planting new woodland). No Income Tax relief is available against capital purchase (including a mortgage or other loan, or annual expenditure costs).

### Inheritance Tax

Commercial woodlands (including both land and timber) qualify for 100 per cent business property relief provided they have been owned for at least two years. Both amenity and commercial woodland can be made a lifetime transfer over seven years.

You may be able to qualify for Inheritance Tax exemption if your woodland is thought to be of outstanding landscape or nature conservation value. Normally this is only considered for very large areas of woodland of national importance that are Sites of Scientific Interest or in National Parks.

### Corporation Tax

Where companies own woodland that is independent of their trading operations, there is no liability to corporation tax in respect of income realized through the sale of growing timber or payments received under forestry grant schemes.

*Lesser celandine (*Ranunculus ficaria*) is one of the first spring plants to bloom.*

### Value Added Tax

Expenditure on forestry operations will normally attract charges for VAT, so it is generally advantageous to register for VAT purposes. The present threshold for compulsory VAT registration is where taxable supplies exceed £68,000; however, voluntary registration is also permitted. Despite the long-term nature of forestry, which often involves long periods of expenditure without receipts, HM Revenue and Customs will normally accept voluntary registration on the basis that there is a firm intention to produce taxable supplies at a future date.

### Stamp Duty Land Tax

Stamp Duty Land Tax is applied to woodland property purchases at a tiered rate according to the value of the transaction.

### Business Rates

There are generally no property taxes (business rates, etc.) on woodland.

## CONCLUSION

So, there is a lot to consider. There are a myriad of rules and regulations but many of them are quite simple to sort out. There are lots of places to find help and a good part of this assistance will be free of charge. Even if you have to pay for some of the advice, this should be regarded as an investment in getting it right, avoiding infringing the rules and in not making expensive mistakes.

# Chapter 4
# Coppice Management

## INTRODUCTION

Coppicing can seem very daunting to a novice. We often have an instinctive feeling that cutting trees down is wrong. But you know, if you have read the preceding chapters, that you can bring many benefits to your wood and the wildlife within it if you overcome your worries and take out your saw. Your aim is to cut almost all the trees and shrubs in your wood off at ground level, leaving only a scattering of trees to grow on as standards. In response, all that mass of energy stored up in the roots of those trees will push forth as new growth transforming your 'clear cut' area into a thicket of coppice poles teeming with birds and bugs and swathed in woodland flowers in just one season. But, how to do it correctly?

## SILVICULTURE

Throughout the life of a broad-leaved tree, buds are produced in the growing layer under the bark. Most of these remain dormant or eventually die off due to the dominance of the buds nearer the top of the tree, which develop to produce the branches and twigs that make up the main tree canopy. These suppressed buds are mainly what produce the coppice shoots when the tree is felled. The act of felling breaks the dormancy and the buds, which may be many years old, burst into life.

Sometimes, regrowth occurs from adventitious buds that are formed on the callous growth formed in the year of cutting, which grows around the edge of the cut stump. These shoots are not particularly common, but can be found especially on sycamore, beech, horse chestnut

*New growth on hazel – only a few days old.*

*Hazel stool cut rather high.*

and lime. This kind of growth can be weak and sometimes doesn't live for more than a few years.

The vigour of coppice regrowth depends on many factors. These include: the age and size of the stool, species, soil, climate and amount of light received by the stool. The relationship between stool or stump size and species is an important one, and although this might not necessarily help much in predicting the number and vigour of coppice shoots, it is a useful way of assessing how many stumps could die, rather than producing new shoots following felling. For example, a birch 30cm (1ft) in diameter is far less likely to be able to survive felling and produce coppice shoots than, say, a birch of only 15-cm diameter or an ash of virtually any size. Some species are very reliable at producing coppice regrowth – these include hazel, alder, ash, small-leaved lime, willow, sweet chestnut, holly, elder, hawthorn. There is a range of species that are normally quite reliable and these include sycamore, hornbeam, oak, elm, aspen, field maple, birch, crab apple, yew, whitebeam and rowan. There are a few that are not really coppice species including wild cherry and beech.

Generally, once a tree has become mature (for the species), the weaker the regrowth is likely to be after felling. Sometimes the tree will produce only a few shoots and these may die; this is probably because the dormant buds have failed or are situated under a thick layer of old bark. If you are coppicing an area that is, say, more than forty years old, you should expect some stump mortality, especially of the weaker coppicing species. Mature birch frequently dies after felling, and strong regrowth from the stump of an oak standard is uncommon. Although a strong coppice species, some of the hazel stools that have been in deep shade for thirty years or more will probably die after cutting. Regrowth will be less vigorous on poor, thin or stony soils, and in shady conditions such as in north eastern facing woods or where the growth is shaded by nearby trees.

## *Cutting Technique and Height of Cut*

As we have seen, the new shoots arise from dormant buds under the bark of the cut stump. Therefore, it is important that you cut the stool to a height that retains some of these buds. On older stools especially, the cut shouldn't be into the older wood, but is better just above the height of the previous cut.

There is evidence (Harmer 2003) that higher-cut stumps produce more shoots – although sometimes shoots arising from these have restricted development at their base due to thick bark and this results in a weak joint that cannot support the vigorous long shoot – this collapses resulting in fewer shoots per stool; this is especially common on birch, willow and ash.

So, on balance and as a rule of thumb, the cut should be as low to the ground as is practical. This is especially applicable to hazel, as this encourages new shoots to arise from close to or even below the ground and thus helps the new coppice stool remain stable. The other benefits of cutting low are that the risk to your ATV/other extraction vehicle is minimized, and also you are far less likely to trip over a low-cut stool. This may seem trivial, but when working in a coppice coupe, it's practical to make the work environment as easy as possible. Historically, stumps were cut low so horses (whether working the wood or on a hunt) didn't injure themselves.

The main exception to cutting as low as possible is where the existing stool is large – it may be a valuable wildlife resource with lichens, mosses, fungi and invertebrates that rely on old wood for a habitat; there are also likely to be nesting holes for birds and small mammals and so large stools should be retained intact. A further reason for not cutting these right to the ground is that sometimes stones can find their way into stools and cutting into these with your chainsaw can mean a lengthy break to re-sharpen the chain.

In some woods, it is clear that stools have been cut relatively high in the past – perhaps 30, 60 or 90cm (1, 2 or 3ft) off the ground. There are a number of theories as to why this was done. Stools might have been cut high so that regrowth is above the height of the browsing attention of brown hares, or perhaps it was a way of marking which local individual or family had cut a particular coupe. None of these theories are very convincing. However, as a historical and cultural artefact, you may like to continue this tradition if it occurs in your wood. Stools cut high later in the twentieth century are sometimes the mark of poorly supervised volunteers or beginners.

Much of the guidance on coppicing states that coppice cuts should be sloping. Some even state that all the cuts should slope to the south. The theory behind this is that a sloping cut sheds water and reduces the chances of fungal infection and, therefore, of rot in the stool. Perhaps a more probable reason is that a sloping stool will also shed sap rising from the cut stump in spring; sap is far more likely to host a fungal infection as it is rich in sugars and other nutrients; it certainly turns quickly into a variety of colours in only a few weeks, probably due to bacterial or early stages of fungal infection.

*Large hazel stool cut just right.*

*Alder stool cut at about 0.7m (2¼ft).*

*Early coppice growth is sometimes suffused with shades of dark red – here on birch; young coppice leaves may be excessively large or of a shape not representative of the mature tree. These birch leaves are deeply indented.*

Following experiments, the Forestry Commission found that there is no evidence that the type of cut (sloping or flat) affects stump mortality, shoot numbers or height growth (Harmer 2003). Of course, over time, the centre of a stool will rot anyway. Most stools contain fungal activity in them, but trees are capable of compartmentalizing this activity, i.e. restricting it to certain areas, creating a boundary between living and dead wood. It feels intuitively right, however, that cuts slope away from the centre of the stool, and this is a common tradition across the whole country.

It has been suggested that where there is luxuriant moss growth around the base of coppice stools, which occurs mainly in western areas, it should be cleared away from the stool to allow the dormant buds to be stimulated by the warm sunshine on the bark (Ted Green, pers comm.). Again, although this seems sensible, there is little evidence that stools covered in moss don't respond to coppicing.

It is good practice to finish the cut stump by making a clean cut to remove evidence of the felling cut; this makes for a more attractive end result and, again, although there seems to be no evidence, probably helps to reduce fungal activity and promotes quicker callusing over of the cut stump. It is important that cuts are clean with no tearing separation of bark from the remaining stool.

Generally, all the trunks on the coppice stool should be cut off. Sometimes coppice workers carry out the practice of going through a coppice (this applies especially to hazel) and taking only the stems from a stool they require for a particular product. It is important not to take more than a shoot or two or you will compromise the ability of the stool to regenerate itself.

Trials have investigated whether there are any differences in coppices cut by hand compared to those cut by chainsaw. In the early days of chainsaw use, traditional coppice-workers were concerned that there were detrimental effects on the coppice. However, no differences were found in the trials. There is something quite spiritual and primitive in cutting coppice by hand, but for those running a business cutting coppice, use of a chainsaw is the only practical option.

### When to Cut

The best time to cut coppice is during the winter when the trees are dormant. There are several advantages to winter cutting. Less foliage makes for easier working, and due to there being less sap in the wood, the bark is less likely to tear. Less sappy wood also has advantages for the greenwood or craft worker in that the wood tends to last longer as a product; this is probably due to the wood being less attractive to decay fungi resulting from it having a lower sugar content. Regrowth will spring from coppice stools in most months of the year; however, any growth beginning from mid-July to September won't have time to set a dormant bud and will be susceptible to winter frost damage; in many species, this will result in a forked or kinked stem.

Winter cutting reduces the potential for disruption to all kinds of wildlife. Most plants are below ground or reduced to their bare minimum; many birds have migrated back to their winter haunts in Africa and others aren't breeding anyway, and many invertebrates are hibernating. If your site has any of the European Protected Species on it, you have a legal obligation to protect it from deliberate or accidental harm and this can influence when coppicing operations can be carried out (see Chapter 3 for more details).

Having said all this, most woods would have been worked traditionally through much of the year; this is still the case in the commercially worked areas of south-east England. The commercial coppice industry here cannot afford to have a protracted period when coppice isn't cut. Although we are very protective of wildlife, and it is inevitable that nesting birds will be disturbed if coppice is cut in May, there is little evidence to suggest that there has been any significant damage done in the past by coppicing in the spring and summer. It does seem likely, however, that there wouldn't have been much coppicing during the months of August and September as in times gone-by, most of the rural population would have been involved in the agricultural harvest.

In summary, most people coppicing today tend to avoid felling trees from the end of March until the end of September. There is an argument to begin coppicing earlier in the season than is traditional and to finish earlier. The sap in many species, especially birch, which comes into leaf early, is running from mid-February, and as we have seen already, there is good reason not to coppice once this has started. In addition, with climate change evident, bluebells are often above the ground in February and it is known that they are vulnerable to trampling damage.

**Benefits of winter cutting**

- Longer-lasting product.
- Less disturbance of breeding birds.
- Less trampling of flowering plants.

However, cutting outside the winter months is OK except for during August and September.

## *Coppice Rotations*

The coppice rotation is the number of years between cuts. Traditionally, the coppice rotation in woodland would depend entirely on the size of the raw materials needed to make the products required. Hazel on a good site would often be managed on a 6-year rotation to produce rods for hurdle-making. In south-east England, hornbeam managed for firewood, would probably have been cut on a 20-year rotation, whilst in the Lake District, a mixed coppice cut for bobbin wood was cut on a 12-year cycle and oak coppice cut for charcoal and bark production was cut on a 20–25-year cycle.

In theory, managing a coppice for a set group of products should be a simple mathematical calculation of dividing the area of coppice in the woodland by the number of years it takes to grow the materials for your products. This gives you the area you need to cut each year to maintain the right rotation. For example, a 5-ha ash coppice managed to produce firewood, might need to be coppiced on a 15-year rotation. This would result in a third of a hectare needing to be cut each year; after 15 years, the coupe you first cut would be cut again. Of course, if you needed more wood than this, you are going to need to cut another wood as well. A 3-ha hazel coppice managed to produce rods for hurdle-making would need to be cut on a rotation of 6 years; this would mean that half a hectare should be cut each year.

However, few people nowadays have the luxury of cutting a managed coppice in rotation with the right number of standards and a set plan for obtaining the raw materials for a set group of products. A neglected or out-of-rotation coppice should be treated quite differently. There are two schools of thought about how to treat a neglected coppice, and your route to its restoration depends on a number of factors, including your management objectives. By coppicing the oldest, most neglected area first, you will be able to stop these areas declining further; of course, you're likely to have larger material to deal with and may have to do more work to make the coppice better for the next cut, for example, gapping up. While you're coppicing these very poor areas, the better areas will continue to decline. The second method would be to cut first the area that hasn't declined in quality quite so much. This means that your raw materials are better quality, you need to spend less effort on making the coupe better for next time, and the results for biodiversity may be better too. The downside to this is that the poorer areas you've not cut will decline further! There is probably no right or wrong answer to this conundrum, and if you can make the choice, the answer is probably to cut some of the better areas as well as some of the more neglected areas.

Of course, this is where the notion of a coppice rotation becomes hard to pin down, and you may only decide to cut an area for the second time, when you feel it is ready; in other words, you decide on the rotation length when you get there. This goes against the concept of proper management planning, but it does have the advantages of being flexible and enabling you to cut the coupe when the material has reached the right size for your products. The most important aspect of coppice rotation is to remember is that the rotation is not complete until canopy closure has been reached. This is crucial as it means that the vigorous vegetation, such as brambles, the grasses and the tall, ruderal species, can die down to nothing, thus allowing the smaller, flowering herbs to be in full sunlight when the coppice is next cut. The clean woodland floor also makes for a much easier working situation for the coppice worker. This full canopy closure could be reached at year five for dense, vigorous hazel coppice but may not be reached until year ten for a neglected mixed coppice.

*Hazel coppice in the fourth summer after coppicing showing bramble growth dying out under its shade.*

### Should Coppice Coupes be Next to Each Other?

In addition to the coppice rotation, another factor that should be considered when working out where to coppice is the relation of one to coupe to another. Some people like to disperse the coupes throughout the wood so that no coupes are cut next to each other in consecutive years; others prefer to coppice adjacent coupes. There seems to be more potential advantages to wildlife to coppice coupes next, or at least very close, to each other. The main benefit is that this is thought to encourage some wildlife to move more easily from one area of young coppice growth to another; clearly, if your new coppice coupes are separated by 200m (656ft) of dense, neglected coppice, this is a barrier to species easily moving from one coupe to another; some insects, such as butterflies, are notoriously poor at colonizing good habitat, even if it is close by; birds on the other hand would be able to find your new coppice much more easily. If you are coppicing as a hobby, perhaps over a few weekends a year, and can only manage a relatively small coupe in one season, to coppice adjacent coupes means that over several seasons, you can create an area large enough to let in sufficient light for strong regrowth; if you can only manage small coupes and you scatter these throughout the wood, the benefits to the wildlife and coppice itself are vastly reduced.

Finally, a further advantage of this method is that, given a choice, deer prefer to browse in quite small areas; therefore, a larger open area (of adjacent coppice coupes) is less attractive to deer. If you are using temporary deer fencing, there may be benefits to working on adjacent coupes because you might be able to reduce the number of fence posts needed or even reduce the length of fence required.

As a rule of thumb, coppice coupes should not be less than about $50m^2$. This equates to a quarter of a hectare or just over half an acre. This

*Hazel and birch being cut in October with 4-year-old coppice in the background within plastic deer fence.*

is because light is such an important factor in the vigour of the coppice regrowth, and full sunlight maximizes this growth. A south-facing coupe can afford to be smaller than a shadier north-facing coupe.

There may be other factors that could influence where your coppice coupes are and the order in which you cut them. In the wood managed by one of the authors, coupes are cut in a sequence designed to reduce the likelihood of windblow and an effort is made not to create a wind-tunnel effect; they are also designed to keep the main yard area screened from the footpath and public highway, as well as with landscape effect as a consideration. In addition, this wood is steep and rocky in places and boggy with small streams in other places. Coupes have to be shaped and situated so that the quad bike can be used to extract the produce, avoiding the steep, rocky and boggy bits, and they have to be cut in such a sequence that the quad isn't driving over new coppice regrowth. All in all, there's a lot to consider.

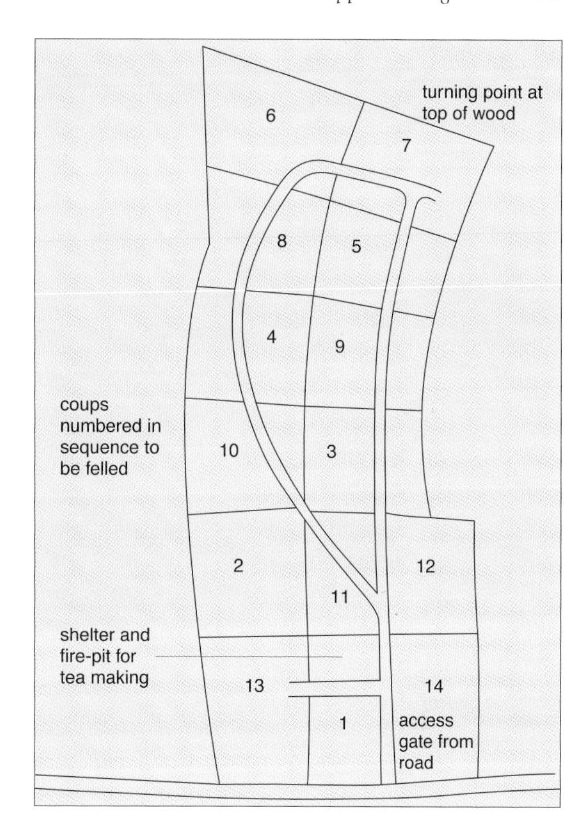


*Map of coppice coupes – in this wood, some coupes are adjacent and some are further apart.*

## *Rides*

In some woods, there will be a network of rides. These can be managed just like linear coppice coupes by cutting the coppice along their length in a suitable (usually fairly short) rotation. This can increase the connectivity within woodland to the great benefit of some species, especially butterflies; at ride junctions, the corners can also be cut to make a larger open glade. This management can be quite intensive but has proven benefits.

### Tips for regenerating old coppice

- Open up as large an area as you can manage.
- Break up these areas into smaller, more manageable coupes at the second cut.
- Introduce interconnecting rides to encourage wildlife to move around the wood and to aid extraction.
- Aim for the standards to cover no more than 30 per cent of the canopy area of the coupe.

## *Standards*

Now to the tricky issue of standards. This is a subject that has frequently confused modern coppice managers, mainly due to the complexity of managing coppice for wildlife, as well as producing a useful product.

In the age before efficient machinery, large trees, although valuable, were time-consuming and difficult to fell, to extract from the woods and to process. Well-grown standards provided useful, medium-sized trees that were needed for construction but were not of such a size that they were difficult to handle. Historically, the evidence is that over 90 per cent of standards were oak and they were usually harvested well before they reached 100 years of age. They provided the kind of material that the coppice growth couldn't.

Historically, the number of standards in a coppice was defined by laws and systems that had been followed for generations. The general

*Old standards can enhance biodiversity – here by providing a nesting or roosting habitat for birds or bats.*

rule of thumb is that the number of standards in a particular age group is half that of the younger age group below it. The mature standards were felled when the underwood had been coppiced, thus removing the oldest generation, and a new generation of standards was recruited by selecting maidens that had set seed at the previous coppicing. This is the reason that coppice that has been managed over the last hundred years or so, rarely has any really old or 'veteran' trees in it.

There has been a wide variety of 'standards per hectare' quoted over the years. Perhaps the most commonly seen is 16 per acre (40 per ha), though the range is from 5–40 per acre (12–100 per ha) (Rackham 2003). In a historic context, it was relatively rare for an oak standard to last beyond 70 years of age before being felled.

Clearly, with the aim to have a variety of ages in your standards, there will be a range of sizes too. This means that one mature oak will take up much more room than a young ash standard, for example. In fact you could get ten young standards to one mature one. This means that, perhaps a better measure of the right number of standards will be to consider the percentage of the canopy that they take up. The rule of thumb here is that in a coppice-with-standards system, about 20–30 per cent of the canopy area should be covered by standards. This could be made up of just a small handful of large oak standards or up to several dozen smaller standards. It is useful to remember that an oak standard will cast relatively heavy shade, whereas a birch or ash standard will cast a lighter, more dappled shade.

Oak has always been the preferred standard species; there are few alternative species that could be used, with ash being the main one; others that are considered acceptable include sweet chestnut, elm, hornbeam and birch. Beech casts too heavy a shade to be used as a standard. Standards are normally grown from seed but, where appropriate seed-grown trees have not emerged with the coppice, it is possible to select single stems from coppice stools to grow on as standards. Timber quality won't be as good, as the trunk is likely to be swept (curved) to some extent at its base and could even become hollow through fungal attack in due course. There is also a chance, if this is done, that the stool will be unstable and could blow over. Our recommendation is that if there is no appropriate standard available, do not contrive to make one.

*Coppice with standards in the winter – note the wide spacing of the standards.*

There are plenty of examples of woods that have been coppiced and inappropriate standards have been retained. The biggest mistake made with selecting standards is that too many are kept. This is especially the case with 'conservation' coppicing. It is tempting to leave shrubs that have a known high wildlife value or produce berries, such as hawthorn; sometimes, trees or shrubs are retained that are either not considered to be traditional coppice species or that may be locally rare, such as crab apple, holly, bird cherry or small-leaved lime. All these species will coppice quite happily. I have seen all trees with honeysuckle growing up them retained. Sometimes, if these individual trees and shrubs are kept standing, you are left with a coppice coupe that is neither one thing nor the other. Too many standards and shrubs will reduce the light reaching the coppice regrowth, making it weaker and more susceptible to deer browsing or being overtaken by vigorous bramble growth; the ground flora won't respond as well and the resultant benefits for birds and insects will be poorer.

### Stool Density

Spacing depends somewhat on the vigour of the stools in your particular wood. An area of big, vigorous stools on good soils can be further apart than an area of weaker stools on poor soils. In a mixed coppice, most species, apart from hazel which we discuss later, will grow best between 3 and 5m (9–16ft) apart. If your plan is to grow your trees as a crop and to maximize yield, then you may want to increase your stocking density. It can be difficult to increase stool density because, if you try to establish new plants in between existing stools, they may not be able to keep up with the fast coppice regrowth of their neighbours and die. If the coppice is very widely spaced, it is possible to plant up the gaps (see the guidance under establishing a new coppice). Any planting should be done in gaps that are sufficiently large to allow in enough light for the new plants to grow well. If you are planting in woodland with a vigorous field layer, it is advisable to mark your planted shrubs so you can find them, to free them of physical competition – a stake with a pink or yellow painted top is quite easy to see. Use of 1.2m (4ft) tree shelters is not advisable, especially where light levels are already low; 0.6m (2ft) tree shelters will help you find your trees easily and offer some protection too, although when you come to stump back, they will be time-consuming to remove and put up again. Plants should not be too big – 45–90cm (18–35in) is the optimum. It is tempting to dig up young plants that are in the wrong place, such as a track, and move them. However, this is also time-consuming and it is much more cost- and time-effective to order suitable plants from a forest nursery. You may want to consider layering in hazel and sweet chestnut coppice.

Of course, some open areas are beneficial to wildlife, provide a sheltered place for deer to feed on grass, not coppice, and, if you are willing to forego a little productivity, some of these areas can even be maintained as being open for that extra bit of biodiversity.

## *Layering*

Hazel has the ability to produce roots quite easily where stems are in contact with the ground. Layering is a traditional technique that takes advantage of this and which you can use to bolster the density of hazel plants. When coppicing your hazel, leave a medium-sized rod on each stool that is close to a gap that you want to fill with a new hazel (you can take two or three off one stool if necessary). When you've finished everything else in the coupe, bend over the rod and work out where it will touch the ground, about 1m (3ft) from the end of the rod. At this point, scrape away the leaf mould on the ground with your boot heel and take off a peeling of bark on the underside of the rod at this point (you can use a knife or billhook). Bend it right over and peg it down where it meets the soil. You can use a simple hazel peg or a heavy weight such as a rock; cover the meeting point with soil. Within a year or two, this should have rooted and will start to produce new vertical shoots, indicating that a new plant has formed. After another year or so, you can remove the parent rod to encourage the new plant's independence.

There are two variations on this theme. In some places, a traditional technique is to heap up soil around the base of a hazel stool to encourage rooting into this soil. After a year or two, the hazel will have produced sufficient rooting for the soil to be scraped away and the hazel coppiced very low so that you end up with rooted rods that can be planted as new hazel stools.

*The layered hazel is pinned to the ground with a forked peg to get a good contact between the living wood and the soil to promote rooting.*

Secondly, a variety of layering was developed not to produce new plants but to create a useful raw material. Some walking-stick makers used to bend over a hazel rod flat to the ground, pegged along its length and still attached to its parent. This has the result of producing vertical rods right along the length of the horizontal rod, growing at a right angle to it. After three or four years, this produces several ready-made walking stick handles.

### *Yield*

A frequently asked question is 'how much wood will I get from my coppice?'. This is usually asked in relation to producing firewood, as it is important to know whether you will get enough wood from a particular area to feed your firewood needs. Unfortunately, the answer is not straightforward. Foresters use a system where tree growth is measured in cubic metres per hectare per year, known as yield class (YC). So, an average ash coppice will put on 6m³ of timber per hectare per year over the life of the coppice cycle. This is a useful place to start and normally forms the basis of our own calculations in the absence of a better system. To complicate matters, different species grow at different rates at different soils in different parts of the country, so this section can only give a very rough guide.

You may well already know that oak grows quite slowly – about YC 2 in the uplands, perhaps 4–6 on better sites; ash is faster – normally YC 6, along with alder and birch; sycamore and cherry up to YC 8 and poplar can reach 18. Most conifers grow faster than broad-leaves; the pines are the slowest, with the firs and spruces the fastest, notably Sitka spruce which can reach YC 26. Even this system assumes certain spacings and stocking density and thinning regimes. Many coppices suitable for producing firewood will be mixed and with a variable density of stools. Strictly speaking the Yield Classes assess the maximum yield, so as a rule of thumb, volumes calculated should be rounded down; there are other complex reasons to do this. Perhaps the best way to have a look at this is with a couple of examples – see the box at the top of the opposite column.

**How to calculate yield of wood from coppice**

**Example 1.** A mixed coppice on poor ground, left neglected for 40 years, might be growing at YC 4. So, 1ha × 40 years × YC 4 = 160m³/ha. You might want to be cautious because there are some wide gaps between stools, or perhaps a corner will be difficult to get to, so a more realistic estimate could be 130m³.

**Example 2.** An ash coppice on better ground, with good stool density, not coppiced for 30 years, growing at YC 6. So, 1ha × 30 years × YC 6 = 180m³/ha.

**Example 3.** A newly planted alder coppice at 2m density, 12 years since stumping-back, growing on very good land growing at YC 8. So 1ha × 12 years × YC 8 = 96m³.
However, at only 12 years old, some of this simply won't be big enough to make saleable firewood. Harvest at year 15.

### *How to Deal with Brash*

Traditionally, very little that was cut was wasted. Everything had a use from the branch wood of the standards to the brash that was tied into faggots or pimps for fuel. The coppice industry had hundreds of years to develop its own local rotations according to the species growing and the local needs of the area. Coppice products were valuable and this led to most coppice being cut at its traditional rotation for the products desired. During the twentieth century, the coppice sector gradually became less vibrant and waste became more commonplace after coppice working, except during hard winters when everything was used for fuel. Once this began to happen, coppice was cut less often, resulting in more waste at each cut. It became traditional in many areas to pile waste or brash into windrows – long lines of brash – and organized working, along the lines of commercial felling of conifers by chainsaw, tends to promote this.

The conservation coppicing of the 1970s and 1980s tended to reduce the waste into great big piles of ash by having enormous fires. Then, during the 1990s, habitat piles came into fashion after it was recognized that large fire sites caused damage to the woodland. Wording in the UK Woodland Assurance Scheme (UKWAS – the standard for best practice woodland management) allows burning provided some lop and top is left for habitat, and provided location and density of fire sites are carefully planned.

The modern prescription for dealing with brash is usually to create habitat piles and to avoid burning. However, in a neglected coppice, there can be so much brash that piles can get truly massive and impossible to manage; large brash piles (and windrows) can mean that a large area of the woodland floor is covered to such an extent that the wonderful ground flora is shaded out. Leaving the brash spread over the site is an option but this can make it difficult (and dangerous) to walk and/or drive over the site. Large fires damage the soil to such an extent that it takes decades to recover, and large fires can damage surrounding coppice stools and standard trees. However, small to medium-sized fires help to deal with the large amount of brash, provide a means of cooking a hot meal and of keeping you warm on those freezing winter days.

The mixture of small fires and some brash piles seems to be the best compromise for managing unwanted brash. Small brash piles provide for some of the nesting and habitat requirements of small birds and mammals, whilst small fires remove the real surplus of brash that gets in the way of efficient working and covers up too much ground flora. Retaining some brash also has the advantage of returning some nutrients to the soil. Fire sites can be ugly, but they soon become covered with plants, such as willow-herb, and after a couple of years can be invisible to the untrained eye. One way of reducing the damage a fire does to the ground is to use corrugated

*Brash windrows keep the work area tidy.*

iron sheets raised up off the ground to support the fire. Using the largest logs that you have felled, the aim is to raise the fire 30cm (1ft) above the ground. Ideally, the ash should then be removed from the site or spread over a wide area to reduce the impact of nutrient release on a small area. Regardless of whether you have a fire on the ground or on a metal sheet, you need to avoid damage to surrounding coppice stools and standing trees.

Chipping has been used as a method of reducing large quantities of brash. However, this can be expensive and, although the woodland floor can absorb a light covering of chips, a pile of chips will quickly kill off the vegetation underneath as the chips decompose and release a large quantity of nutrients back into the ground; if you're going to chip, it's probably best to remove the chips from the wood.

**Tips on dealing with brash**

- If there is a lot of waste, burn brash on fires raised off the ground, or on areas of little value such as rhododendron stumps.
- If there is not much waste, scatter it around and it will rot away very quickly.
- Make brash piles but keep them small and dense.
- Consider chipping but remove the chips and compost elsewhere.
- Making dead-hedging keeps brash relatively tidy and in one place, and helps to deter deer if tall enough (a dead hedge is really just a tidy wind-row).

## DEER

Any book on coppicing needs a sub-chapter on deer. The presence of deer, the species and population level, will influence virtually every woodland-management decision you take. The problem is that deer eat trees! If you don't protect your coppice regrowth, they will eat it, as

*Roe deer are probably the most likely species you will see and are especially at home in the coppice.*
*(Photo: Teresa Morris)*

*Red stag; red deer are the largest species of deer in the UK.* (Photo: Teresa Morris)

*Fallow buck.* (Photo: Teresa Morris)

well as any natural regeneration, and sometimes, many of the wild flowers too.

The number of deer species in Britain has been augmented over the years by introductions. The only species native to Britain are red deer (*Cervus elaphus*) and roe deer (*Capreolus capreolus*), with fallow (*Dama dama*), muntjac (*Muntiacus reevsii*), sika (*Cervus nippon*) and Chinese water deer (*Hydropotes inermis*) all being introduced. All species are widening their range, although Chinese water deer and sika are still relatively restricted.

So, if coppicing is such a traditional management technique, how was regrowth assured, when there have always been deer in the British countryside? The answer to this often asked question is twofold. Firstly, there were more people living in the countryside and they looked after their coppice woods. The coppices were full of valuable produce and so the woods were protected with banks and ditches, as well as, in some cases, either palisade fencing or laid hedges. There were usually people living near the coppices and deer could easily be moved on or shot. The second reason is food. It's easy to forget that before the mid-twentieth century there were no supermarkets; the once weekly shopping trip to the nearest town, usually on market day, was quite an expedition. Bad weather frequently meant that roads were washed away or people were snowed in where they were. This meant that people were often hungry and actively hunted deer as a source of meat, whether they had permission or not, and so the deer population was far smaller than it is now. The population explosion of deer has come about since food became cheaper in the decades after the Second World War, and rural people didn't need venison on their menu to survive.

There are many people who would like to live in harmony with all living things, and would not entertain the possibility that deer would have to be shot. However, the reality is that humans have been interfering with nature for so long that further intervention will always be needed. Deer no longer have a natural predator in Britain, except perhaps the motor car. People in the countryside don't need to hunt for food any more, and deer are thriving. This is encouraged by these times of climate warming where mortality of deer in the winter is lower and the grass-growing season is longer. There are thousands of hectares of woodland that people rarely visit and these provide a quiet sanctuary for deer. So, how can you ensure that your coppice will grow?

There are several options to consider. Firstly, the optimum situation is to have a deer population balanced with the habitat in which it is living, so that there is no necessity for any kind of deer fencing. Inevitably, this involves culling. This is not all bad! Although stalkers are clearly shooting deer for a number of reasons – sometimes for pleasure and sometimes for a living – the great majority of them have the interests of the deer at the top of their agenda. If a deer is to be shot, it is of supreme importance to them that it is despatched with a shot that causes instant death.

Clearly, if you allow stalking to take place in your wood, you should ensure that the stalker has the relevant certificates and insurance, as well as the necessary experience to carry out the job effectively. The Level One Trained Hunter qualification is essential. The welfare of the animal should be of paramount importance. An experienced stalker will not necessarily shoot a deer on every visit to the wood, but will wait to take the right individual at the right time. There is clear evidence that stalking carried out well results in a deer population that is healthier and probably happier. The advantage to stalking your wood efficiently is that the requirement for expensive deer fencing is reduced or eliminated; in addition to this, you gain access to a source of lean, tasty and very healthy meat. There may be situations where stalking is not your top option, although I would urge that it is considered, even if you wouldn't eat the resulting venison.

So, whilst understanding that a balance between the animals and their habitat is the optimum situation, in many circumstances fencing may be the next best option. It is important to bear in mind that permanent exclusion of deer from an area has implications on the vegetation, both within and outside the fence. By erecting a fence to exclude deer, whilst carrying out no culling, you will be putting more pressure on surrounding habitat, especially if the exclosure is large. Vegetation within the fence will also change; as well as unbrowsed coppice growth and good natural regeneration, brambles will thrive and could begin to invade any open ground, including paths and tracks; any holly regeneration will begin to grow quickly and could soon dominate some areas.

There are several fencing options, as shown in the table on page 66.

*Plastic deer fencing is relatively cheap, very flexible and can be used several times.*

## Advantages and disadvantages of different types of deer fencing

| Type of deer fencing | Advantages | Disadvantages |
|---|---|---|
| Permanent<br>High-tensile wire | Extremely effective.<br>Excellent recruitment of natural regeneration between existing stools.<br>No browsing of woodland flora. | Cannot be moved to a new area.<br>Expensive.<br>Once a deer is in, it can be difficult to remove or kill.<br>Brambles can get out of hand. |
| Temporary<br>Plastic mesh<br>Galvanized wire | Much cheaper.<br>Easy to erect.<br>Can be re-used.<br>Good recruitment of natural regeneration between stools.<br>No browsing of woodland flora. | May be less effective in some areas with high muntjac or red deer populations.<br>Brambles can get out of hand, unless moved by year 3. |
| Electric | Relatively cheap. | Easily becomes ineffective without constant monitoring.<br>Not suitable for long periods.<br>Requires power source. |
| Chestnut paling | Effective in small circular patches.<br>Can be re-used. | Very heavy to move about.<br>Only works for small areas. |
| Brash hedges around stools | Can be cheap in small areas.<br>Allows bramble to be browsed.<br>Visually natural.<br>Utilizes a waste product.<br>No materials cost. | May break down unless construction is good.<br>Limited or no recruitment of natural regeneration between stools.<br>Limited effectiveness in muntjac areas.<br>Very time-consuming to erect.<br>Not enough brash on second-cut coupes.<br>Labour intensive. |
| Brash hedges around coupes | Some recruitment of natural regeneration between stools.<br>Bramble can be browsed once the hedge breaks down. | Time-consuming to construct.<br>Limited effectiveness in muntjac areas.<br>Not enough brash on second-cut coupes. |
| Wigwams | No bramble problem. | Doesn't work unless deer population is very low anyway. |

The best fencing option for you depends on the exact situation. A high deer population level will mean more pressure on the chosen fence. A common mistake is to try to protect too large an area in one go. A permanent fence of larger than, say, 2 or 3ha will be hard to keep deer-free, especially if there is public access through the compartment or a water course. Fences on steep, rocky ground can be hard to make completely stockproof. In addition, once a deer is in, it will be almost impossible to either get it out or to shoot it.

Many people will be tempted to avoid fencing and to use brash piles, stacks or wigwams over each coppice stool. These are only a viable protection method if the pile of brash is well-constructed or simply enormous! Wigwams are the wrong shapes to protect regrowth from the attentions of deer, which just push their noses in to eat the luscious young shoots. Brash fences around each stool are fine if they are well-made and large enough; a roe deer has a long neck and, as the coppice shoots grow, they spread out and so the construction needs to be at least 1m (3ft) away from the stool in each direction. In practice, these are time-consuming to erect and frequently break down before the regrowth is out of harm's way.

Our favourite is the use of temporary plastic fencing. A 50-m roll can be carried by one person and erected in half an hour without the need for stakes every few metres. It can be tied up on to existing trees with cheap baler twine; sometimes it needs a hanging wire from which to suspend it between trees. It can be kept in place for two or three growing seasons, depending on the rate of growth and then moved to the next coupe; it is extremely tough and can be re-used many times. The black type is not as ugly to look at as you might think, especially once honeysuckle has grown up it, although this can make taking it down rather difficult!

Deer are magnificent animals and no-one wants to see them eradicated, but there needs to be a serious national push to reduce the population, especially of non-native species, and particularly of muntjac. This is a small species capable of squeezing through very small fence gaps, which not only eats coppice regrowth but can do severe damage to our wild flora by eating flower heads – the oxlip in Suffolk has been badly affected. The Deer Initiative may have a field officer in your area to help with bespoke advice.

*Permanent deer fencing is expensive and inflexible.*

## PESTS, PROBLEMS AND WORRIES

### *Sycamore and Beech*

Sycamore is considered not to be native to the British Isles. Beech is not considered generally to be native north of a line from Dorset to Norfolk, although there is evidence that it spread north of this to become established as scattered colonies as far as Yorkshire and Cumbria. Both species are naturalized, i.e. they will regenerate from seed in the wild. Sycamore in particular has a bad reputation amongst some conservationists as being a weed and a tree to be destroyed at all costs. When it grows in the right place, foresters tend to like it as it produces a clean, white timber, which is usually in great demand and the better quality timber can be exported.

Neither species has been used traditionally for standards. Beech, in particular, casts a very heavy shade under which little can survive, including hazel coppice. So, as a rule of thumb, it should be removed from most coppices as a young tree. Sycamore is perhaps a better candidate as a standard and it won't necessarily invade woodland by seed, although it enjoys establishing itself on disturbed ground. It is probably more of a threat on richer, alkaline soils; where it grows as a continuous canopy, it also casts a heavy shade. There is some evidence that sycamore is of benefit to dormice, probably due to its propensity to harbour massive numbers of aphids.

*Grey squirrel damage on 6-year-old sycamore coppice.*

### *Grey Squirrels*

This species was introduced to the UK in the early twentieth century and it has spread to almost all corners of the country. Although many people like to see these animals – they are entertaining to watch in your garden – they have caused havoc in many of our woodlands and are detested by foresters. As if that wasn't enough, the grey squirrel carries a virus that kills the native red squirrel, it strips the bark of some broad-leaved trees, predates bird eggs and devours most of the hazel nuts produced each year. The bark-stripping activity isn't well understood but it is thought that it is something to do with males marking territory. Damage occurs mainly on the smooth-barked tree species, especially beech and sycamore. It can be so bad that large parts of the canopy are ring-barked – these branches then die and the canopy begins to fall apart. Grey squirrels should be controlled, either by shooting or trapping. In an actively managed coppice, grey squirrels tend not to be too bad a problem, although they can certainly badly damage young sycamore coppice.

### *Honey Fungus*

Most people will be familiar with this fungus as a garden pest. It is one of the few fungi that will aggressively attack living trees, although this is relatively rare and normally confined to smaller, exotic ornamentals. Most trees, especially in a woodland situation, will not be affected by honey fungus unless they are heavily stressed by another agent such as drought, damage by fire or perhaps an insect pest (or dead). The spores of this and other fungi are everywhere, so there is nothing you can do to control it anyway.

*A large windblown ash over a public footpath.*

### *Windblow*

This is normally an issue reserved for the more exposed north and west of the United Kingdom, although, of course, there are some storms such as that of October 1987, which have the ability to cause a lot of destruction anywhere. Often, there is little you can do about the threat of trees blowing over. Perhaps the most important management tools are to retain a wind-firm edge to your coppice coupes and not to create a wind-tunnel effect in the direction of the prevailing wind – normally the west or south-west. Windblow is a natural event and creates excellent habitat for a range of biodiversity, especially on up-turned root plates. Even if you harvest some of your blown-over wood and timber, it is easy to leave plenty for wildlife, and possible to create some very valuable additional habitat niches.

### *Conifers*

Many conservationists have put all conifers into one easy category – they are bad. It is not as simple as this though, and there is often benefit in retaining some conifers. None of the conifers are native, except yew and the Scots pine in Scotland, and this is the reason they are disliked. It is certainly true that, if successfully planted in ancient semi-natural woodland, they are likely to kill off a large part of the coppice species – shrubs and ground flora – through deep shade and needle fall. If this is the case, though, the recommendation for managing these woods (sometimes known as Plantations on Ancient Woodland Sites – PAWS), is to remove the conifers slowly to maintain the high humidity and shelter, but to allow in sufficient light to stimulate any ancient woodland remnant features to become more robust and begin to flower. There should be no compulsion to eradicate all the conifers from a wood, unless there is a real threat of them taking over to exclude other species and, in most circumstances, there will be benefit to biodiversity by long-term retention of some individual conifers. Perhaps the only exception to this is regarding western hemlock (*Thuja plicata*), which has the ability to grow in deep shade and can therefore eventually produce dense thickets and exclude almost everything else.

## PLANTING A NEW COPPICE

It is relatively straightforward to create a new coppice wood. A newly established woodland will lack most of the interesting features of a much older one but can at least provide a supply of raw materials for craft use or for fuel.

In order to establish a new coppice, all the basic rules of creating any new woodland apply. There isn't room to cover this in any detail here, but there are some points specific to establishing your new coppice that we should cover. First of all, the new wood will need to be composed of the kind of coppicing species that will provide you with the materials you want. Most people want to establish a new coppice to supply fuel. If this is the case, you need to plant fast-growing species and the best species for this are probably in the following order:

1. Ash.
2. Sycamore.
3. Silver birch.
4. Alder.
5. Hazel

Eucalyptus is a sixth option. It is not native, so you wouldn't get the wildlife benefits of the other species, but it is very fast-growing and provides a good, dense firewood. You could plant a mixture, but what does best will probably depend on the soil and other environmental conditions. It is vital to weed your trees, if you want quick establishment. Many people feel they have failed if they resort to use of herbicides to control grass and weed growth, and this is understandable. However, three years of keeping vigorous grass growth under control using a well-tried and tested herbicide, such as glyphosate once a year, will do very little to harm wildlife. If your new coppice is established in a field formerly under agriculture, the land quite possibly received three chemical applications per year for the past thirty years! Rapid establishment of the woodland

Eucalyptus gunnii *on the left with oak and birch at the right, Nottinghamshire.*
*(Photo: Andrew Leslie)*

### Tips for creating a new coppice wood

- Plant no more than 2m apart.
- Choose a fast-growing species, such as ash, birch, alder or sycamore.
- Protect against damage from animals.
- Weed with a herbicide for at least three years.
- Stump back after three years to obtain multi-stemmed trees.

*Hazel seedling in its first year of growth.*

means that you will get new trees that can be coppiced sooner. This is also aided by planting at a high density. Planting at the maximum 3m allowed under the recent grant schemes is far too wide for a new coppice. Many would argue that 2m is also too far apart and spacing as close as 1.5m can give your coppice a head start. The new trees can be stumped back (i.e. cut back to a stump with a pair of secateurs) after one or two growing seasons when they are evidently well established; this will make them multi-stemmed.

Following stumping back, new woods of the appropriate species could be successfully coppiced for the first time between the ages of roughly five and twenty five. If such a wood has been planted with grant aid within the last ten years, it would be sensible to inform the grant-aiding body that you are coppicing the wood and not grubbing it out. As with an established coppice, regrowth will need to be protected against damage from grazing animals. A new wood less than five years old is still very early in its establishment phase and by coppicing it, it is in some ways set back. In order to get the best growth, the trees need to reach the 'canopy closure' stage as soon as possible. This draws the stems upwards, resulting in straighter growth, but it also means that less light reaches the woodland floor and it is this that shades out the competitive grass growth. A young wood that is still quite grassy will benefit from an application of herbicide or a good mulching to reduce this grass competition and encourage canopy closure quickly.

## CONCLUSION

If you are not sure about taking a saw to your trees, then don't hesitate to attend a couple of courses, or pick some of the Royal Forestry Society's summer visits to attend (after becoming a member of course). Ask the Forestry Commission woodland officer for advice about felling licences. There are lots of ways to get help. There are very few absolute do's and don'ts. The most important thing to remember is that as long as there are no grazing or browsing animals, and there is enough light, your coppice will grow. Some trees will die but new ones will take their place, provided animals don't eat them. You must be certain that the coppice will not be eaten off once the new shoots appear; and, if they do get eaten, then you should be prepared to take action by culling deer or fencing off the coppice.

# Chapter 5
# Nature's Abundance

## INTRODUCTION

The aspect of coppicing that never ceases to amaze all those who have experienced it is the effect the sunlight has on the coppiced area. Not only do the coppiced trees grow again but the dormant plants flower, seeds germinate, new trees begin, brambles sprout and the wood becomes full of birds and butterflies. This chapter looks at the plants and animals that characterize coppice woodland from the largest to the smallest.

*Hazel catkins are a sure sign that spring is on its way.*

## THE COPPICE CYCLE

In Chapter 4, we described the pattern and variation of light and shade in a managed coppice rotation. Much of the abundant biodiversity of a coppice is evident in the few years after cutting, but there are plants that can survive for many years – sometimes decades – under shade. However, there aren't many that actually need these conditions to thrive (the rare violet helleborine (*Epipactis purpurata*) is one possible example), although there is some evidence to suggest that wild garlic (*Allium ursinum*), dog's mercury (*Mercuralis perennis*), and herb paris (*Paris quadrifolia*) are set back by removal of overhead shade. Most plants need at least some sunshine to flower and either set seed or to be vigorous enough to spread by vegetative reproduction. Common dog violets (*Viola riviniana*), for example, can survive for over thirty years in the deep shade of a neglected coppice or on track sides in ancient woods planted up with conifers; sometimes they have so few very dark green and violet-shaded leaves that they are difficult to spot; they only flower in full sunlight.

The beauty of a regular coppice cycle is that the sunshine and warmth that many plants need comes around on a regular basis; in many cases, by the time of the coppice being cut, plants still have some vigour from the previous coppicing and their flowering can be profuse. For generations, coppices have been renowned for their carpets of woodland flowers. Coppice coupes are frequently sheltered by surrounding uncut coppice making them up to 3°C warmer than neighbouring shady woodland (Buckley 1992). The coppice with standards habitat niches are also less exposed to drying winds and hard frosts and so provide a sheltered environment in more ways than one. A wood with an active coppice rotation will have coupes in a variety of stages and this provides a great length of woodland 'edge' habitat. A lot of wildlife prefers woodland edge where the open areas are warm and sheltered with flowering plants, whilst close by is dense cover providing areas to shelter and rest.

The flowering of plants is just one aspect of the extraordinary resurgence of wildlife that enjoys the warm, sunny conditions. Insects take advantage of the nectar and some butterfly species rely particularly on coppice coupes to thrive. There are many bird species that enjoy the habitat provided by coppice coupes. We'll expand on this profusion of wildlife during the course of this chapter.

*Primroses flower profusely after years in the shade.*

**Some species that thrive in a coppice rotation contrasted with those that don't**

| Type of wildlife | Examples | Wildlife usually adversely affected by coppicing | Examples |
|---|---|---|---|
| Butterflies | Most fritillaries, skippers. | Some butterflies. | Purple emperor, green-veined white. |
| Invertebrates that feed on flowering plants | Bees, hoverflies. | Invertebrates that require lots of deadwood. | Beetles. |
| Firespot fungi | | Fungi that require continuity of large deadwood. | |
| Most ancient woodland plants | Too many to mention. | Plants that require a degree of shade. | Violet helleborine, herb paris, bluebell. |
| Small mammals | Dormouse, voles and shrews. | None. | |
| Large mammals | Deer. | None. | |
| Birds | Warblers, nightingale, robin, wren, spotted flycatcher, woodcock, tree pipit. | Birds that prefer open woodland with scattered trees and hole nesting birds that need ancient trees. | Redstart, pied fly catcher, some tits, woodpeckers. |

## ANCIENT WOODLAND AND COPPICE

Ancient woodlands are defined as those being in existence at around AD1600. The theory is that people only rarely planted trees before then (and they planted woodland even more rarely), so if a wood can be said to exist at around that time, the probability is that it has existed since the land was invaded by trees after the last Ice Age, perhaps 10,000 years ago. In medieval times, most woodland would have been coppiced (more especially in England than Scotland or Wales but not exclusively so) and it is therefore likely that most ancient woodlands would have been coppiced, either just a few times or many times for generation after generation. It is known, for example, that the Bradfield Woods in Suffolk have been coppiced continuously to this day since 1252 (Rackham 1976); if they were coppiced on a 15-year rotation, these woods have probably been coppiced at least fifty times. Having said this, not all ancient woods are coppice woods, and not all coppices have ancient woodland origins.

The importance of Ancient Semi-Natural Woodlands (ASNW) cannot be under-estimated. The continuation of woodland on a site for thousands of years means that these sites have a combination of habitats and ecological niches that cannot be replicated. The soils will have rarely been disturbed to any great extent and the associations that have developed over the millennia between trees, shrubs, invertebrates, fungi and plants, such as orchids, are pretty much impossible to replace artificially. A newly planted wood will take thousands of years to develop a biodiversity resource that comes close to that of an ancient wood. Ancient woodlands are truly irreplaceable and have been described as being akin to our medieval cathedrals in their importance to our country.

There are a number of ways to find out whether your woodland is ancient. Natural England holds provisional inventories of Ancient Woodland, and this information is publicly available either from Natural England or on the MAGIC website; it is quite easy to look up whether your wood is ancient or not. However, the inventory continues to be called 'provisional' and isn't perfect, so you may want to do

*Ancient semi-natural woodlands are often characterized by a sinuous woodbank, especially in southern England.*

some of your own research. Try to find your wood on old maps – you will probably have to go earlier than the ordnance survey first edition and look at more locally made maps produced before around 1840. You might be able to research local records, such as estate maps and plans, wills or auctions and sales to find references to your wood at an earlier date, but this could be painstaking work.

There are clues to look out for in the wood itself. In southern and central (mainly lowland) England, the presence of a sinuous wood-bank and ditch are often a sure sign of ancient origins; there is a wide variety of ditch and bank designs and combinations. The woodland name itself could be a sign of its origins, although this is fraught with difficulties to catch out the unwary. Ancient woodland names could include those ending in '–ley', for example (from the old English 'leah' meaning 'wood' (Rackham 1986), whilst non-ancient origins include the names 'plantation', 'furze' or 'cover'. You may be lucky enough to find some large coppice stools or boundary pollards, which could indicate that the wood is ancient, but it takes an experienced eye to interpret these features. The absence of old trees does not mean the woodland is of relatively recent origin.

Ancient woodland tends to be situated on land that was difficult to cultivate (before the age of large, modern machinery). In the uplands, this means that if your wood is steep or boulder-strewn this is probably a sign that it has never been converted for agricultural use. In the lowlands, woods that have not been cultivated before are often on the steep sides of river valleys or hillsides, or they are on heavy clays forming plateaux that are frequently waterlogged. In south-east England, ancient woodlands are often at the edge of the parish, sometimes with the parish boundary also being the woodland boundary. Some woods have ponds, walls or other boundary features, such as ditches, within them. These features aren't necessarily a sign that the wood was once fields – sometimes people constructed boundaries within woods. Ponds can be natural – perhaps the result of a glacial feature or man-made – sometimes the result of digging for clay.

*Toothwort is an ancient woodland indicator plant, here growing with two other AWIs – ramsons and sweet woodruff.*

Finally, another sign that your wood could have ancient origins is the presence of ancient woodland indicator plants (AWIs). These plants spread very slowly usually due to poor seed dispersal abilities and, therefore, they colonize new woodland at a slow rate. AWI plants are those that are generally found in ancient woodland only. This is a massive generalization, and there is a very wide geographical variation of what can be considered a reliable AWI. As a rule of thumb, there are more reliable AWIs in the south and east of Britain than there are in the more oceanic climate of the north and west. AWIs tend to be tolerant of long periods of shade but, along with most other plants, they flower profusely when opened up under a coppice regime.

Of course, the presence of one AWI doesn't necessarily mean that your wood is ancient; a judgment needs to be made of a combination of all the evidence above plus a suite of AWIs to determine the likelihood of the wood's ancient status. Perhaps the undisputed king of ancient woodland indicator plants is the herb paris, which has one large black seed that simply drops off the plant and has no obvious means of dispersal.

*Herb paris is probably the ultimate ancient woodland indicator plant species.*

*Bluebell seeds spread at a rate of about 1m (3ft) per year.*

**Some of the more common and strong ancient woodland indicator plants**

Wood anemone (*Anemone nemorosa*); Sweet woodruff (*Galium odoratum*); Water avens (*Geum rivale*); Yellow archangel (*Lamiastrum galeobdolon*); Common cow-wheat (*Melampyrum pratense*); Yellow pimpernel (*Lysimachia nemorum*); Dog's mercury (*Mercurialis perennis*); Early purple orchid (*Orchis mascula*); Sanicle (*Sanicula europaea*); Wood speedwell (*Veronica montana*); Common dog violet (*Viola riviniana*); Moschatel (*Adoxa moschatellina*); Lily-of-the-valley (*Convallaria majalis*); Remote sedge (*Carex remota*); Toothwort (*Lathraea squamaria*).

### *Seed Banks*

The flora of a coppice wood seems to comprise two distinct types, which have developed alongside each other (Buckley 1992). There are the forest species, which are adapted to persist in their vegetative state in shady conditions; these tend not to produce a seed bank because of this ability to survive in deep shade for many years. Then there are those light-demanding species, which cannot survive as living plants during the shady periods but produce seed that can survive until the conditions are suitable for germination. The length of time these seeds can survive depends on a variety of factors, such as how much seed is eaten by woodland creatures before germination, soil acidity and so on. It is thought that after about fifty years of shade, the amount of seed that survives in the soil reduces rapidly; so, if your coppice has been neglected for longer than this, or if it was planted up with conifers more than fifty years ago, the woodland flora will probably take much longer to recover.

Most seed-bank plants appear in the first or second year after coppicing, and they last between two and five years depending on the density and vigour of the stools. Seed-bank plants include bramble (*Rubus fruticosus*), raspberry (*Rubus idaeus*), the St. John's worts (*Hypericum* sp.), figwort (*Scrophularia nodosa*), wood spurge (*Euphorbia amygdaloides*) and ragged robin (*Lychnis flos-cuculi*). Foxglove (*Digitalis purpuraea*) is a species whose seed is known to be able to persist in the soil in a viable state for decades; it readily germinates and forms an abundant component of the coppice flora on the more acidic soils. Rackham (2003) mentions broom (*Cytisus scoparius*) as being a species that comes in after coppicing. There are undoubtedly other plants that do not have an obvious means of dispersal that can survive as

*Water avens is a strong ancient woodland indicator – here with cuckoo spit.*

*The foxglove has seeds that can last up to 100 years in the soil and they readily germinate on the disturbed ground of a freshly cut coppice.*

buried seed for perhaps up to 100 years. Some woods have a particularly characteristic coppice flora that gives them an identity all of their own.

### Bare Ground

Coppice woodland suffering a long period of neglect will gradually lose its ground flora, at least for a time. This often means that when neglected coppice is cut again for the first time in many years, there is sometimes virtually no ground vegetation. This bare ground can be the perfect seedbed for plants, and the trampling and dragging of brash can enhance the conditions for seed germination. These areas often attract 'ruderal' species, i.e. plants that invade a habitat on a temporary basis – most garden weeds are ruderal. In a coppice, this may include some familiar species, such as field pansy (*Viola arvensis*), groundsel (*Senecio vulgaris*), dandelion (*Taraxacum officinale*) and ragwort (*Senecio jacobea*). A non-native that has become a pest in some woods is buddleia (*Buddleja davidii*), which readily invades bare sites. These species are light-demanders, however, and they don't usually persist for more than a few years as the coppice regrowth out-competes them and they disappear until the next coppicing. Of course, there are many plants that will grow in these woodland conditions but they are not necessarily characteristic of them – they are just adapted to making the most of temporary niches.

A tree species that readily colonizes bare and disturbed ground is birch. In mixed coppices, this forms a fast growing, useful component; however, where it invades a coppice grown for a specific crop, such as hazel, it is sometimes considered to be a pest to the extent that it is weeded out. In a derelict coppice with large gaps between stools, birch can be very helpful by filling in the gaps and helping to close the canopy early, suppressing the brambles and drawing the other coppice stems upwards; it casts a relatively light shade and, except where it is extremely dense, it can co-exist quite well with most other species.

Another tree species that likes to seed itself onto disturbed ground is sycamore. This tree has had a bad press over the years due to its propensity to set huge quantities of viable seed that germinate to form dense stands of seedlings, especially on richer, alkaline soils. It casts a heavy shade and it's definitely a species that you don't want in a hazel coppice, as it will heavily suppress hazel growth to the extent of killing it. However, sycamore will coppice quite well, provides good firewood on a relatively short rotation and is probably a better supporter of biodiversity than most people give it credit for.

*Scorpion flies (*Panorpa hyemalis*) mainly eat dead insects and thrive in a young coppice as soon as there is any shade.*

### Deadwood

Much has been written (and assumed) about deadwood in coppices, and there is a presumption that coppices are a deadwood desert.

Clearly, there will be certain invertebrates that need deep shade and a certain kind and quality of deadwood, perhaps reminiscent of the kind of wildwood image that most of us have imagined with massive decaying tree hulks. Most coppices won't possess this kind of deadwood and won't be suitable for these species. Sometimes, a coppice coupe can look extremely bare, especially after over-zealous tidying up, and this cannot be good for creating a rich and biodiverse wood. So, retention and creation of deadwood is an important component of modern coppice management. There are four types of deadwood to consider.

The first is existing lying deadwood. Clearly, deadwood already on the woodland floor is more useful to wildlife, at least initially, than anything you create. This is best left in situ or, if it's in the way, just moved to one side; you can mix it with other habitat piles you are creating. It should definitely not be burned or taken home to the woodshed.

Secondly, you can create deadwood from the coppice you cut. This is easier with the first cut of a neglected coppice as the size of the wood is larger than if the coppice is in a managed rotation. Freshly cut wood can be left scattered on the ground, in piles or stacks in the sun or shade, on its own or mixed with existing deadwood. Probably the two most important things are to create a variety of deadwood in all kinds of situation and to leave some really big bits of wood that take a long time to decay. If all your deadwood is birch, it's likely to have disappeared by the time you come to cut again, but if some is oak with heartwood, this could be rotting away over several coppice rotations.

If you are lucky enough to have a watercourse in the wood you are looking after, it is very tempting to clean out these rivers and streams by removing branches and logs. However, this is all part of the natural processes of nutrient

*It is quite easy to create spectacular deadwood like this by ring-barking large trees such as sycamore and letting them die and crash to the ground.*
*(Photo: Alan Shepley)*

*Deadwood in water courses increases the amount of available food downstream.*

re-cycling in woodland. Deadwood in water courses catches leaves and other vegetation, which form the nutritional basis of the lower end of the food chain. If it's cleared out, the biodiversity will be poorer as a result. If watercourses are cleared out on a larger scale, this increases the risk of flooding too.

A third variation of deadwood is standing deadwood, i.e. dead trees that are still standing – small ones are sometimes called 'spires'. As a rule, these should be left in situ provided they are not a danger if they fall. The bigger the standing deadwood is, the better it is likely to be for wildlife; species like birch will rot quickly and provide a great woodpecker home for a few years before they eventually collapse. Species like oak and beech will remain standing, sometimes for decades, and provide habitat to more specialist species. Standing deadwood is especially important for some species of bat. Unfortunately, leaving spires can make the area you've worked look rather messy and ugly, but you can be reassured that retaining these is the best thing you can do for biodiversity.

You can create your own standing deadwood by ring-barking, i.e. cutting into the bark and the live wood underneath right around the circumference of the tree close to its base. This cuts off the supply of water and nutrients from the roots to the canopy. Often, trees take 2, 3 or even 4 years to die, but this can be speeded up by painting the cut surfaces of the ring-barking with glyphosate. For obvious reasons, it's important not to create dead trees next to any rights of way or other hazards. These trees will gradually collapse once they have died, and provide an important habitat for bats, hole nesting birds, small mammals, invertebrates and fungi.

The final kind of deadwood is brash. For centuries, there was almost certainly no waste resulting from working a coppice coupe – everything would have been used. Early conservationists restoring neglected coppice usually burned the resulting brash but towards the final couple of decades of the twentieth century, as the knowledge of the benefits of deadwood grew, the conservation movement took to substituting burning with actively creating habitat piles of the unwanted brash. Brash rots very quickly, and in many cases you won't be able to tell where the brash pile was when the next coppicing is done. These piles of twigs and branches provide good nesting places for small birds, such as wrens and robins, and are useful for invertebrates and some fungi; they will also provide good protected hunting areas for small mammals.

Although fire sites damage the soil underneath and destroy the immediate seed bank, they do provide habitat for nearly fifty species of fungi that depend on fire sites for a place to live – the so-called 'phoenicoid' species (Spooner and Roberts 2005). So fires are not all bad. These areas of black ash can also be the best place for reptiles to bask in early spring sunshine, as they warm up very quickly.

### Rides and Glades

It is not easy to define what a woodland ride is. Essentially it is a woodland access track, though a ride tends to be less formal than a conventional forestry track, and can have wide margins that are sometimes mown or cut on a rotation of a few years. A glade is a semi-permanent open area within woodland; it may remain open for several reasons, such as dense bracken, deer browsing or mowing to enhance conservation. Earlier in this chapter, the concept of woodland edges was introduced – rides and glades provide a kind of permanent woodland edge feature within a woodland. Good ride management is crucial to maximizing their wildlife benefit. Originally woodland rides would have been made to allow easy access for extracting woodland produce and later for management of game birds and stalking deer. A real bonus is that they provide sheltered, semi-permanent open areas, often with herb-rich grassland. It is known that butterflies can use rides to travel from one part of a wood to another – this means that they could relatively easily find a coppice coupe of the appropriate age and structure by travelling along these rides. Of course, the rides need to link up and be relatively wide to provide the best benefit. It is thought that east–west orientated rides are better than north–south rides, as they are sunnier and, therefore, warmer.

Good ride management should produce a good edge structure with bays cut in the woody vegetation like mini-coppice compartments, and the herb layer should also ideally consist of a variety of different growth stages. This kind of management takes up resources in terms of time and machinery, and so you have to be dedicated to improving the wildlife of your wood to do this – or else have a pheasant enterprise, as they will also benefit from this management.

*Good ride structure.*

*High brown fritillary (*Argynnis adippe*) has suffered a drastic decline over recent decades, part of which is thought to be due to changes in woodland management, including a reduction in coppicing.*

*Nut-tree tussock (*Colocasia coryli*) is a common woodland moth. (Photo: Colin Reader)*

## THE WILDLIFE

### *Butterflies and Moths*

One of the most serious declines in Britain's biodiversity during the twentieth century was that of woodland butterfly populations. The butterflies that have come off worst are those needing open habitat within woodlands – this was once provided by coppicing. Formerly, coppice woods were full of species like the silver washed fritillary (*Argynnis paphia*), pearl-bordered fritillary (*Clossiana euphrosyne*), Duke of Burgundy (*Hamearis lucina*), wood white (*Leptidea sinapis*), heath fritillary (*Mellicta athalia*), purple emperor (*Apatura iris*) and white admiral (*Limenitis camilla*); the blues and skippers also like open sunny rides and glades. Some of these butterflies have an incredibly complex life-cycle and it's not surprising that their habitat has been disrupted so much by our activities.

These days, you'll be lucky to find any of the butterflies mentioned above in your coppice, but in parts of central southern England, the Morecambe Bay limestones and one or two other favoured spots, you might be lucky. However, other species can take advantage of the warmth and shelter of today's coppice coupes, such as orange tip (*Anthocharis cardamines*), brimstone (*Gonepteryx rhamni*), specked wood (*Parage aegeria*), small tortoiseshell (*Aglais urticae*), meadow brown (*Maniola jurtina*), green-veined white (*Artogeia napi*) and ringlet (*Aphantopus hyperantus*).

Although we know far less about moths due to their mainly nocturnal habits, there are at least twenty times as many moth species as butterflies. It is thought that about half of common woodland moths are declining. Just like butterflies, moths can have a very complex ecology, and surprisingly little is known about many species, such as what their food plant is. Moths need flowers or fruit on which to feed and places to hide during the day, such as dense vegetation, a variety of bark niches or piles of brash or leaves. Their larvae (caterpillars) need suitable host plants (virtually every woodland plant plays host to a moth species) and when the time comes to become a pupa, they need the right kind of vegetation, piles of leaves and other woodland debris, and the right soil conditions in which to pupate.

However, although moths are complicated creatures, it has been shown that open rides and glades are critical to the survival of many moths in the same way that they are vital for butterflies. It has been suggested that moths prefer more mature coppice than most butterfly species (Ellis 2006); this makes sense as there would be all the right habitat requirements close to hand, with dense shade, high humidity and a good layer of leaf litter, but some sunny spots with flowering plants too.

In some years, one or two moth species can produce massive numbers of caterpillars. On a still day in May or June, it is sometimes possible to hear the caterpillar droppings falling onto the vegetation below the canopy. Some small birds time their eggs to hatch around the peak population of this insect life and in these years of caterpillar abundance, they must do very well.

### *Birds*

Perhaps one of Britain's most iconic and best-loved birds, the nightingale (Luscinia luscinia), has suffered one of the most severe declines in its population due to loss of suitable habitat. The nightingale is a bird of dense coppice regrowth, although it will live quite happily in a variety of dense scrub habitats too. Although it is a migrant species, it seems to have a limited ability to colonize rejuvenated coppice, tending only to return to places where there has been a long continuity of coppice management. Clearly, this bird's relationship with both its wintering habitat in Africa and its nesting habitat in Britain is a complex one and, currently, we appear to be able to do very little to halt its decline. There are other bird species that prefer to nest in young coppice (younger than ten years of age), including some that have suffered sharp declines. This list of declining birds that use coppice for nesting and/or feeding, includes: turtle dove (*Streptopelia turtur*), dunnock (*Prunella modularis*), song thrush (*Turdus philomelos*), lesser whitethroat (*Sylvia curruca*), willow warbler (*Phylloscopus trochilus*), spotted flycatcher (*Muscicapa striata*), willow tit (*Parus montanus*), linnet (*Carduelis cannabina*), bull finch (*Pyrrhula pyrrhula*) and yellowhammer (*Emberiza leucocephalos*). In the wood coppiced by one of the authors in the Lake District, another declining bird, woodcock (*Scolopax rusticola*), overwinter and breed in mixed coppice under ten years old.

Many of these birds nest on the ground or within the dense field layer of vegetation close to the ground, or in the nooks and crannies formed by the older coppice stools. Research by the British Trust for Ornithology has shown that in order to encourage the bird species to breed in new coppice, it is not enough to simply cut the old coppice – the growth has to be protected from browsing by deer. If deer are allowed to browse heavily, they reduce the bramble growth and the density of other plants, (as well as browsing coppice regrowth itself) and the necessary cover for small birds is not adequate to encourage nesting. This is a clear indication that if you want to encourage birds that prefer cop-

*The nightingale's dramatic decline over recent decades is partly due to the reduction in the amount of coppiced woodland.* (Photo: www.rspb-images.com)

*The marsh tit (*Poecile palustris*) is now a 'red-listed' bird due to the steep decline in its population.* (Photo: www.rspb-images.com)

pice and you are in an area with a significant deer population, you may well need to invest in some kind of deer fencing or control.

Research has also indicated that many of these birds (like many butterflies) enjoy the varied habitat provided by a mosaic of mainly small coppice coupes. This goes back to the 'edge' effect that was mentioned earlier in this chapter. Obviously, each species will have its own slightly different preference when it comes to woodland type, coppice density, species, age and coupe size. Nightingales, for example, prefer coppice coupes of over a quarter of a hectare, whereas willow warblers and black caps (*Sylvia atricapilla*) are quite happy in much smaller coupes.

One of the habitat niches missing in active coppices that are in rotation is nesting holes for birds. These areas tend not to have old trees or much standing deadwood. For some species, this can be remedied by providing bird boxes; blue tit (*Parus caerulus*), great tit (*Parus major*), coal tit (*Periparus ater*), tree sparrow (*Passer montanus*), nuthatch (*Sitta europaea*), robin (*Erithacus rubecula*), pied wagtail (*Motacilla alba yarrellii*), greater and lesser spotted woodpecker (*Dendrocopos major* and *D. minor*) and wren (*Troglodytes troglodytes*) will all successfully take to bird boxes.

### Large Mammals

It is now centuries since people hunted to extinction many of the larger beasts that roamed our woods and forests. Those that are long gone include bear, wolf, lynx, boar, elk and beaver. One of these, the boar, has escaped from captivity and is now living wild in some woods in southern England; its chances of spreading seem quite good, as it is an elusive and adaptable creature. With large areas of woodland continuing to be neglected, it will have plenty of opportunity to expand its range. Beaver have already been reintroduced and there are moves in Scotland to re-introduce the wolf.

Of course, none of these are especially animals of the coppice, and it could be said that there are no large mammals that rely specifically on a coppice-management regime. You may well

*Leaping red deer stag.* (Photo: Teresa Morris)

be able to watch badgers, weasels and foxes in your coppice, and if you're really lucky and in the right part of the country, you might sometimes spot a pine marten. However, the one animal that is especially happy feeding in a coppice coupe is deer. Everyone is happy to see deer in Britain's woodlands and they are, after all, the largest remaining animal in Britain. However, the coppice worker would probably be the first to say that the population of deer needs to be in balance with the woodland habitat itself, and currently in many parts of the country, there are too many deer. Please refer to Chapter 4 for more on deer in coppice.

### *Small Mammals*

The classic hazel coppice small mammal species is another of Britain's best-loved creatures – the dormouse (*Muscardinus avellanarius*). These animals are incredibly elusive – many people working in coppice only see a handful of dormice in a lifetime spent in the woods. They are mainly restricted to ancient woodlands and hedgerows, and are not known in Scotland or Ireland; they are a protected species. The dormouse is one of the few British mammals that hibernate, and this is probably how it came by its name – in some places it is still called the 'dozing' mouse. It is known that they prefer dense underwood with good areas of fruiting hazel, with flowering plants and shrubs that provide berries to supplement their nutty diet. They are adapted to climbing about in dense young coppice and honeysuckle. In many areas, they have suffered a drastic decline in population, although some re-introductions, along with suitable coppice restoration have been successful.

Dormice have a distinctive way of opening hazel nuts leaving a smooth, round opening, sometimes with just a few teeth marks on the nut surface close to the edge of the hole through which they eat the nut kernel. Finding these nuts is one of the best ways to discover whether you have dormice in your wood. An alternative is to erect specially designed dormouse boxes or tubes, but you need a high density (and a high number) to attract them; the boxes have the hole at the back close to, and facing, the trunk of the tree. If you actually discover a dormouse

*Dormice prefer dense hazel coppice with a good field layer of other plants. (Photo: Michael Woods, the Mammal Society)*

in a box, they are unmistakable, but more often you'll find an old nest (or a blue tit's nest). It can be difficult to tell whether a nest has been made by a dormouse or another small mammal; one very good way is to look at the nest materials. Dormice tend to collect whole hazel leaves while they are still green and these can be found in most dormouse nests, which are roughly round and grapefruit-sized, with no obvious entrance; other mice tend to use fallen leaves that are brown and curly; dormouse nests can sometimes have stripped honeysuckle bark too. If your coppice habitat is in an area where there are known to be dormice, or there are historical records, you may well need to alter your coppice management accordingly to be sure of complying with the European Protected Species Directive (see Chapter 2). This might be by adjusting your coppice season slightly or changing the amount you coppice in a particular part of your wood. Once you know you have them in your woodland, you need a licence to handle them, or even to check whether the boxes are being used by dormice.

Dormice like aerial routeways, i.e. they prefer to be able to move around the coppice without needing to travel on the ground. For this reason, many people have thought that the optimum rotation for dormice was up to twenty years. In a hazel coppice, this rotation is too long to produce good materials and is rarely economic to cut. Of course, for centuries dormice have fared perfectly well in hazel coppiced every six or seven years. The hazel nuts are an important source of autumn food and so at least some fruiting hazel is needed – it tends to start producing nuts at around six or seven years old, and so some hazel needs to be retained to fruit for a few years before cutting.

British woodlands, including coppices, are home to other small mammals too, although there are few that depend quite so much as dormice on coppice. Although not as elusive as the dormouse, you may have to be very patient and do quite a lot of rummaging around amongst the undergrowth to see other small mammals. None of them are confined to coppices; indeed, most can be found in non-woodland habitat, but generally, coppices provide a good habitat for them with their dense and varied layers of vegetation, large number of flowering and fruiting plants and their attendant insects. Wood mouse (*Apodemus sylvaticus*) and common shrew (*Sorex araneus*) are quite common and will colonize a new coppice as soon as ground cover is re-established. The bank vole (*Clethrionomys glareolus*) is quite at home in a coppice and the field vole (*Microtus agrestis*) and pygmy shrew (*Sorex minutus*) also find good habitat when the coppice is a few years old and becomes dense. The yellow-necked mouse (*Apodemus flavicollis*) has a restricted distribution in southern England, but can also be found in coppice habitat. There is probably little you can do to attract these species to your woodland if they're not already there. However, you can help them develop a vibrant population by providing plenty of nesting areas like old coppice stools and pollards, standing dead wood and hollow trees, large fallen deadwood and brash piles; in rocky woods, suitable nesting places can be found under boulders and behind curtains of mosses at the base of rocks and old standards. Supplementary nesting places can be provided with nest boxes – boxes placed for dormouse are frequently used by wood mouse, yellow-necked mouse and common shrew.

There are fifteen species of bat (*Chiroptera* spp.) native to Britain. None could be said to be reliant

*Woodmouse using a dormouse nest box.*

*This old oak standard provides a good home for some bat species in a hole where a branch used to be.*
*(Photo: Alan Shepley)*

on coppice. However, coppice coupes will provide good hunting grounds for bats as they are sheltered and well-populated with insect life, such as moths and beetles that feed on the nectar of the coppice flowers. Most of the bat species need roosting and nesting places in trees. In the same way that you can encourage stronger populations of other small mammals by providing these places, bats will also take advantage of artificially created homes. The minimum required is to retain existing veteran and standing dead trees. You can also create additional dead trees by ring-barking. There are a few woodland owners who have experimented with creating very specific bat habitats in the canopy of large trees. This needs highly skilled chainsaw operators who can use the nose of the chainsaw to burrow into the tree to create ideal nooks and crannies similar to the natural ones found in veteran trees. Please don't try this yourself unless you are an experienced chainsaw user and have the required tree-climbing certificates. Like the dormouse, all bat species are protected by law. Felling a tree that is known to be a roost for bats is illegal and could attract a hefty fine.

### *Reptiles and Amphibians*

There are no reptiles or amphibians that depend on coppice as a habitat, but you may well find grass snakes (*Natrix natrix*) and adders (*Vipera berus*) sunning themselves on the bare ground of a new coppice coupe. You don't need a pond in your wood for it to be a habitat for frogs (*Rana temporaria*) and toads (*Bufo bufo*). Toads in particular like the cool, dark conditions provided by log piles, and these are of course home to one of their favourite foods – slugs.

### *Invertebrates*

If you sit still in a coppice of two or three years of age and just watch carefully, the full range of wildlife will slowly become apparent. In some ways, the bugs and beetles of the coppice are easier to find and watch than some of the birds and small mammals. You often don't need to rummage around in the undergrowth, but simply be patient and observant. If you become really curious you can use some of the entomol-

*Violet ground beetles (*Carabus violaceus*) live under leaf litter and wood piles.*

*This stinkhorn fungus (*Phallus impudicus*) has attracted plenty of flies to feed on it and help spread its spores.*

Rhagium mordax *doesn't have a common name; it is frequent in many woodlands during May and June.*

ogist's methods, such as sweeping a very fine-meshed net through the undergrowth to see what you can catch, or to lay out a white sheet on the ground before beating the shrub layer to see what falls out.

Your coppice will be home to a simply enormous variety of insects – most won't rely on coppice as a habitat but will be perfectly at home there. Scorpion flies, hoverflies, beetles, weevils, bumble bees, wasps, lacewings, gnats, mosquitoes, robber flies, slugs, snails, crickets and groundhoppers are just a few of the creatures you'll be able to see.

These small creatures live in a wide variety of places and some have very exacting requirements. This is why it is beneficial to ensure the biggest range of habitat possible. It is the range in structure, age, species (a wide range but mainly native) and deadwood that increases the range of habitat niches. Having the deadwood standing and fallen, in the open and in the shade, piled up and scattered, in water courses, small and large and of all species enhances the sheer biodiversity there may be.

## CONCLUSION

The biodiversity of a coppice woodland in rotation can be awe-inspiring. It is unlikely that there will be many rarities, but what coppices lack in rare species, they often make up for in sheer weight of biomass. There have been so many arguments put forward in recent years in favour of non-intervention. Britain already has plenty of non-intervention and low-intensity management woodlands. What many woodlands need, in particular woodlands that still exhibit coppice characteristics and especially many ancient semi-natural woodlands, is to be coppiced again. Even if we don't truly understand all the reasons for the decline in some wildlife, such as woodland butterflies and birds, the least we can do is provide the kind of habitat that they require in this country to be successful. If enough woodland is coppiced, we may be able to boost the number of woodland birds and butterflies to make their populations more secure again.

# Chapter 6
# Hazel Coppice

## INTRODUCTION

Hazel (*Corylus avellana*) has been grown as a coppice crop for at least the past thousand years. From the use of wattle in early houses to the trackways of coppiced poles found in the peat bogs of Somerset, evidence can be found for the extensive use of this ubiquitous woodland shrub. Coppiced on a short cycle of 5–10 years, the versatile hazel has been used since ancient times for a range of coppice crafts and products, the most well-known being the hazel or wattle hurdle.

The climax of the hazel coppice industry occurred in the late Middle Ages when the sheep industry took off on the South Downs and other upland areas of Southern England, mainly in the counties of Hampshire, Dorset and Wiltshire.

*Hazel that is overstood will eventually die out.*

Hazel hurdles were in huge demand for folding and controlling sheep before the first of the Enclosures Acts in 1720 changed farming practice, resulting in the planting of many miles of hedges (Rackham 2006). The value of wool and the demand for hurdles were so high that some of the best farm land was planted up with 'simple coppice' (pure hazel plantations)

These woods are in some cases still being worked today and, if you are lucky enough to have access to these productive woods before they disappear, they are a resource to be valued and nurtured. Most of us though will be dealing with more neglected woodlands. These are characterized by old derelict coppice with many mature trees or 'standards, where the standard density is too high and the hazel has become 'overstood' or shaded out and is in danger of disappearing altogether.

By the mid-twentieth century many of the markets for hazel products disappeared. Sheep hurdles were replaced by metal hurdles and fencing, the advent of cardboard and plastics replaced wooden crates and bobbins. In some areas, coppicing as a woodland-management practice very nearly disappeared. There are many hectares of derelict coppice across Britain that have not been cut for a number of decades. These neglected woods suffer a reduction in biodiversity because they lose the multilayered structure and periodic bursts of sunlight of an actively managed coppice.

The past thirty years has seen a revival in interest in coppice management, primarily for conservation purposes, but also in the commercial uses of coppice products. There has been a resurgence of traditional crafts, such as hurdle-making and the development of new products; for instance, rods for straw-bale building (they

*Hazel trellis peeled and painted with microporous paint.*

are used to pin the bales together), rods for geodesic domes or poles for the construction of Mongolian tents or 'yurts'. Craftspeople have responded to the increased demand for natural garden products by developing many variations of trellis and plant supports. The trellis shown in the photograph above is made of hazel rods peeled in the spring when the sap was flowing, dried out and then smoothed down and painted with a microporous paint, which should extend the life of the trellis by several years.

## COPPICE WITH STANDARDS

The majority of the woods worked by the authors in the North-West of England would be considered 'coppice with standards', a silvicultural system where some trees are allowed to mature (the standards), with an 'understorey' of coppiced hazel and other species. There are three key issues for a successful coppice with standards regime:

- Standard density.
- Animal browsing.
- Coppice stool density.

It is essential to attend to all three elements when restoring old coppice.

### *Standard Density*

Most overstood coppices will have far too many standards. The easiest way to gauge the density is to head into the woodland, away from the paths or tracks, and look up into the canopy. For maximum impact do this in mid-summer when the leaves are at their best. You are looking for glimpses of open sky and making a judgement about the percentage of sky to canopy. In an overstood coppice or a wood that has been managed as high forest, the chances are that there will be 100 per cent canopy cover, with all the crowns of the trees touching and no spaces between. For optimum coppice re-growth you should aim for 20–30 per cent canopy cover (about one-quarter canopy to three-quarters sky). See also Chapter 3 for standard density. The percentage scheme is easy to apply when you are familiar with it and is useful for gauging the need to fell more trees as the standards mature.

### *Animal Browsing*

Covered in detail in Chapter 4, but it is worth reiterating here, that for hazel coppice especially, deer browsing must be eliminated. Just one stray deer can do immense damage with a casual bite here and a bite there. The hazel that had potential to grow into a straight flawless

*100 per cent canopy cover – all the tree tops touching.*

*50 per cent canopy cover – half and half tree tops and sky.*

rod, perfect for countless crafts and uses has, in an instant, been reduced to a doglegged, forked stick that might make a peastick if you are lucky. Rabbits and farm stock can also do damage but they are easier to control than our ever-increasing wild deer herds. Culling is needed to control numbers, but even the most assiduous stalker cannot be on duty at all times. Brashing up stools and dead hedges can be effective but is labour-intensive and uses up potential product.

If you are serious about getting a crop of good rods from your hazel, then a deer fence is the only option. It does not have to be a permanent fence but deer-height netting attached to existing trees with string will be the minimum protection required. This temporary fencing has the benefit of being able to be moved on after the hazel has grown above the deer head height (two years) and put up again around a new coupe.

*Hazel that has had a good brash protection and is growing quite well.*

*Deer height netting being put up using trees as supports.*

*Using a post tied with string to protect the tree.*

## *Hazel Stool Density*

In derelict coppice the chances are that the under-storey of shrub species may have already been reduced through mortality. This may be due to the age of the trees but most likely it is because of lack of light. Species such as hazel, contrary to received wisdom, cannot cope with total shade. It is possible to find 50-year-old hazel stools growing in full light that are still throwing up new shoots from the base demonstrating their natural propensity to coppice.

*Hazel regenerating in the light created by the fallen old stems.*

Left to its own devices, old stems would die and break down leaving a gap in the canopy for these new shoots to grow up in its place. These shoots are known as 'sun' or 'summer shoots', of which more later. A dense canopy of standards will drastically reduce the number of these sun shoots and inevitably, without this natural regeneration, old hazel stools will die. This is made much worse where the few shoots put up by old hazel plants are eaten by deer.

Neglected woods will become dominated by plants that are more shade- and browsing-tolerant, such as holly and ivy, and the ground flora will be mainly mosses and ferns. As high forest develops there will be less and less shrub layer as the years go by. If, in your assessment of the woodland (*see* Chapter 2), you have decided that it is appropriate to restore your woodland to hazel coppice with standards, then the chances are that you will have to replace and replenish the hazel in order to get the density right.

The table below describes a system of grading hazel that takes the number of stools per ha and calculates the number of usable stems per stool it would take to achieve the different grades of hazel coppice. Usable stems are considered suitable for hurdle-making, i.e. not forked, branched, damaged or too big (4cm/1.5in being about as thick as is practical to use).

Larger, older but otherwise healthy stools will produce more usable stems per stool and so can be spaced a little further apart than, say, a newly planted coppice where the spacing could be as close as 1.5m (5ft) in order to allow for some losses.

The need for as high a density of stools as possible is not just about getting the maximum number of rods per hectare, but is important in order for the hazel stools and rods to compete with each other and draw each other up to the light, thereby growing longer, straighter rods. Hazel growing out in the open, in isolation from

**System for grading hazel as devised by Howe (1995)**

| Stool density (ha) | Spacing | Grade 1 Number of usable rods per stool | Grade 2 Number of usable rods per stool | Grade 3 Number of usable rods per stool | Grade 4 Number of usable rods per stool |
|---|---|---|---|---|---|
| 1000 | 3m × 3m (10ft × 10ft) | 30 | 20 | 10 | <10 |
| 1500 | 2.6m × 2.6m (8.5ft × 8.5ft) | 22 | 15 | 8 | <8 |
| 2500 | 2m × 2m (6.5ft × 6.5ft) | 16 | 11 | 5 | <5 |

*Bushy hazel growing out in the open.*

*Stunted trees growing in the gaps in limestone pavement (grikes).*

its neighbours, is a much more shrubby and branching bush than the same stool growing in close competition.

Replenishment of the hazel stock can be achieved by layering the young shoots (*see* Chapter 4) or replanting.

## *Other Factors*

One of the problems we face as woodland managers is that the woods that have survived millennia of clearance for agriculture and industrial development are often those very woods on the most marginal of land – steep, inaccessible hillsides, shallow soils with rocky outcrops or boggy land. All in all what we have left is often a far cry from those heady heydays of hazel coppice, when the best agricultural land was planted up to meet the demand for hazel hurdles.

### Soils

Hazel prefers well-drained reasonably fertile, moderately acid to basic soils, such as loam. Water-logged or peaty soil will not grow good hazel.

Soil depth is perhaps the more important factor. Shallow soils where the root run is restricted, or shales that are prone to drying out in drought conditions, will restrict the growth height of your hazel.

In the Morecambe Bay area, many woods are on limestone pavement. A fascinating landscape of exposed rock slabs called 'clints' and deep chasms or 'grikes' where rare and specialist plants lurk but the trees that grow up out of the grikes are bonsai'd from lack of free root run. This is not an ideal set of conditions for good hazel growth. What we find is that between the rocky outcrops lie channels of deeper soils typically glacial 'loess', where the hazel can reach twice the height of its nutrient-starved neighbours. To make the best of circumstances, wherever possible, always favour the deeper soil sites.

The gradient can affect soil depth if the soils are constantly being washed away down hill. A good shrub layer and ground cover is vital to counteract this effect. Re-introducing a coppice cycle has definite benefits in these circumstances, as the rejuvenated roots of the hazel bind the loose soils and prevent erosion. If the site is too steep, though, ease of working and

**To hazel coppice or not?**

Have a good objective look at what is growing on your site already. If there is no hazel, it may be for a very good reason, If there is hazel but it is stunted, then the chances are it will not grow tall and straight once it has been cut. If, however, it has reached a good height (over 3m/10ft) and there are obvious signs of vigour in the stools (shoots coming up from the base), then the chances are it will coppice well.

safety can be compromised, not to mention the logistics of getting your products out.

Hazel occurs naturally up to altitudes of 700m (2,300ft) but inevitably growth will be reduced at the upper limits of its natural distribution.

Bill Hogarth, a coppice merchant from the South Lakes district, had a theory that the aspect of your wood was important: if it faced north, the hazel was more likely to have a spiral growth form (something that only becomes apparent when you split your hazel down the middle and the resulting halves resemble corkscrews). His theory was that the hazel contorted itself following the sun's passage.

## ROTATIONS

Hazel is sometimes coppiced on rotations as short as four years for a product such as thatching spars, which only needs quite slim rods, although this is not considered good practice as it would potentially exhaust the stools in time. For hurdles the ideal rotation would be 6 –9 years, as beyond that the rods become too thick. At 10 years you might get hedging stakes, but in general after that it would be considered out of rotation and in need of restoration. Short cycles

*Nine-year-old hazel showing good suppression of ground cover.*

*Hazel coppice felled and lying in drifts.*

can potentially compromise the benefits of coppicing for biodiversity. For the cycle to be complete there must be canopy closure before the hazel is cut again. This ensures that the 'weed species', such as bramble and coarse grasses, are shaded out and suppressed, and the shade-loving plants, such as mosses, can develop.

Delay in canopy closure may be caused by insufficient stocking of the coppice wood. The closer the hazel is planted, the earlier this canopy closure can occur. At present the tendency for conservation coppicing is to leave the hazel to grow beyond what is commercially viable to recut in order to achieve canopy closure. However, if attention is paid both to stocking densities and to browsing problems, then both nature conservation and commercial aims can be met.

## CUTTING THE HAZEL

- Tools you will need: chainsaw with full protective clothing.
- Hand-tools: bill hook or axe, bowsaw, loppers, string.

Ideally, one should coppice a coupe of at least 0.4ha (1 acre) each year; this allows plenty of light into the coppice and minimizes shading from neighbouring trees. You can get away with less if it is on an edge of the wood or bordering a wide ride. Cutting can be done in one go or just enough to work up in the time available. Felling is considerably quicker than sorting out all the produce, especially if you are using a chainsaw to do the cutting. One tank full of fuel in a standard saw (or less than an hour's felling with a chainsaw) will be enough for a day's dressing out. What is important though is that you take time to get the felled hazel to lie in one direction. It is not always that easy, and you may need to stop and physically move it to one side. This will save you lots of time later, though, when you come to work it up and are not faced with a tangled mess!

## DRESSING OUT HAZEL

Start at the point that you finished felling. The idea is that all the poles will have dropped down in sequence like dominos and you should not need to pull at material trapped under other stems, but always be picking up the top layer. You will struggle at first to find space to put your sorted poles, but as you pick up the pole decide what its use is and dress out (trim neatly) all the side branches. A bill hook is the best tool for this job. This needs to be sharp – the best type has a bit of a hook on the end to pick the rods up, which will help prevent straining your back.

*Correct way to dress out hazel.*

> **Bill hook safety**
>
> Always cut away from your body, either by pushing away in front of you or by bracing the rod against your leg and severing the rod with a blow aimed behind you.
>
> If using the hook to lift a rod, avoid pulling too hard – if it is caught in brambles for instance – as you are then aiming the hook towards yourself and it may suddenly give way, resulting in an accident.

Keep the bill hook flat against the rod and slide it away from you, this severs the side shoots cleanly and closely to the rod with the minimum of effort. Always do this trimming from the base of the rod up towards the sky, as it will make the cleanest cuts.

If you are making peasticks, leave the branches on, but cut to length with a clean cut at about 1.5m (5ft). This severing cut can be made down towards the ground, as this will leave a clean spike on the end of the stick for pushing into the ground.

Keep a bowsaw to hand to cut thicker rods to length. Bundle your produce as you go: place a length of string on the ground with a slip knot on the end and place your rods on top, keeping a rough count as you are working; then, when you have enough,you can tie them up and stack them out of your way. Beanpoles are often bundled in elevens rather than tens, as you then have one extra to go across the top.

Two pieces of string, one at the bottom and one two-thirds of the way up, is usually sufficient. Aim to only handle the produce once. Making a pile of dressed out poles and then going back through them to sort them into products is time-wasting and demoralizing. Be decisive and place each rod in its correct bundle as you work and you will be surprised how quickly your stacks grow.

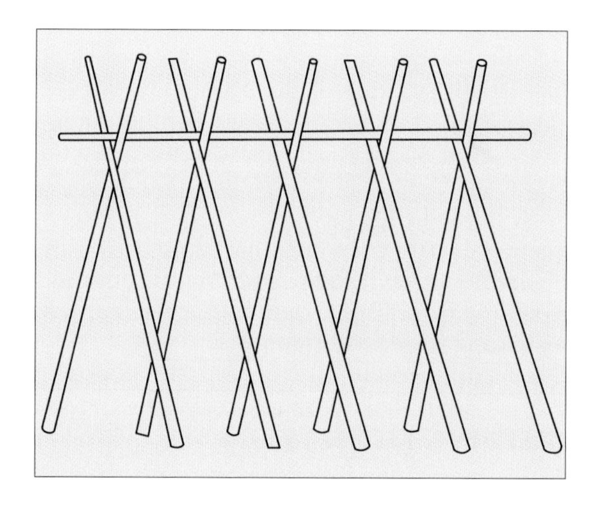

*A row of beanpoles.*

*Sorting and bundling produce.*

## PRODUCTS

At the end of his working life, Bill Hogarth supplied 100 different products, many of which were hazel. These products, including some of our own additions are listed in the box below.

### Coppice products

Hurdles (wattle and gate).
Hedging stakes.
Pea sticks.
Beanpoles.
Ethering (for binding the top of a newly laid hedge).
Strawbale pins.
Swill bools (rims for oak swill baskets).
Net stakes (for sea fishing in estuaries).
Thatching spars.
Scout staffs.
Walking sticks.
Clothes props.
Charcoal.
Pot-pourri curls.
Iron-age reconstruction.
Wattle and daub.
Bundles of brushwood for jumps and sea defences.
Packing rods (for transporting trees in transit).
Besom handles (birch brooms).
Rake handles.
Fence posts.
Garden stakes.
Firewood.
Clothes pegs.
Rustic furniture.
Baskets.
Artefacts for floral decoration (mini hurdles, besoms, etc.).
Chipping for mulch or fuel.
'Gypsy' flowers.
Plant stands.
Geodesic domes.
Stick chairs.
Barrel hoops.
Hurdles.
Woven panels.

What you decide to produce from the raw material will be dictated by demand. Be aware as you work of what products you can sell. It is tempting to make stacks of peasticks, which use up the brashy tops, but if you can't use or sell them all, they will go to waste. In contrast, some rods may be used for more than one product, such as hurdle rods and beanpoles, so you will have to decide which has the most pressing need or is easier to sell. Some products have a high value, like a nice walking stick, and some have a low value, so do remember to save the high-value products first, as long as you have a use for them. The old adage 'a bird in the hand is worth two in the bush' comes to mind here. In the end you should gather the products that you know you can sell. Having a range of products that you are collecting though does mean you can maximize the uses for your wood and minimize waste. On the other hand it is tricky to keep more than half a dozen products in your mind at one time, so don't make it too complicated for yourself.

### *Extraction of Products*

When you have worked your area of coppice and stood back to take a photo of all the fruits of your labours, it is too late to consider how you are going to get all your lovely products out of the wood and back to your workshop.

Transporting bundles out of the wood is labour-intensive and one of the reasons why traditionally much of the value-added work, such as hurdle-making and thatching-spar making, happened within the wood. A good system of rides is a great advantage and an off-road vehicle is very helpful, be it a 4×4, a tractor or quad bike and trailer. If you have access to snigging ponies, this may provide the most environmentally friendly solution.

## ADDING VALUE

### *Hazel Hurdles*

These woven wooden panels are a triumph of ancient skill and technology. Strong and light enough to be portable, they were at one time made by the thousand for agricultural use.

**Hazel products**

| Product | Number in bundle | Ideal dimensions | Price range |
|---|---|---|---|
| Peasticks | 20 | 5ft (1.5m) long, fan shaped with plenty of horizontal branches. | £4–£6 |
| Hurdle rods | 25 (sometimes 20) | 6ft (1.8m) minimum up to 12ft (3.5m) from very thin to 1.5in (4cm). | £8–£10 |
| Hedge stakes | 10 | 5ft (1.5m) long 1–2in (2.5–5cm). | £3.50–£5 |
| Thatching spars | 250 | 2–2.5ft (60–75cm) often cleft from quarters of a 2in (5cm) rod and pointed at both ends. | £20 + |
| Beanpoles | 10 or 11 | 7ft long, 1–1.5in (2.5–4cm). | £5–£7 |
| Besom handles | 20 | 3.5ft (1m) long, up to 1in (2.5cm) thick and smaller for children's brooms. | £10 |
| Strawbale pins | 10 | 4ft (1.2m) long, up to 1.5in (4cm) wide. | £6–£8 |
| Strawbale staples | 20 | 3ft (1m) long, up to 1in (2.5cm). | £8–£10 |
| Yurt poles (walls) | 10 | Up to 6ft (1.8m) long and 1–1.25in (2.5–3.5cm) diameter unpeeled or peeled. | £10–£20 |
| Walking sticks | single | 3–5ft (90–150cm). | £1–£5 |
| Clothes props | single | 6–8ft (180–240cm) forked at the end and approx. 2–3in (5–8cm) diameter. | £5 + |

Traditional sheep hurdles are made on a 6ft (180cm) mould and have ten 'sales' or uprights; the even number meant that a gap or 'twilly hole' could be left in the centre of the weaving. The twilly hole allowed a number of hurdles to be strung on a pole and carried on the shepherd's back to the hillside to make a pen. Garden hurdles are also usually 6ft wide (180cm) but they usually have nine sales, which makes the weave just a little bit looser and a little bit easier to do. They are fixed to permanent posts and used for screens, shelters and a more decorative alternative to the ubiquitous fence panel. Hurdle-making is a tough craft to get proficient at. At times it can seem like a titanic struggle to shape, split, twist, bend and beat the hazel rods into a regular shape, but making a half-decent hurdle is not beyond the capabilities of a fit and active person. To make a good hurdle at a speed necessary to make a living, however, takes a lot of practice.

Not for nothing did the old hurdle-makers hold the apprentices back and had them practicing splitting for years before they were allowed

*Products waiting to be extracted.*

*BHMAT apprentice Sam with a first hurdle.*

to progress to actually weaving. Each operation has to be fast and efficient with a perfect economy of movement. The best way to achieve this is to get very well-organized before you begin.

You will need:

- Mould to hold the uprights or sales.
- Chopping block.
- Rod horse.
- Splitting post.
- Bill hook.
- Bowsaw.
- Loppers (a modern addition and not strictly necessary).
- Secateurs (ditto).
- Bundles of hazel.

## Mould

The mould can be an old beam, sleeper or, if you prefer, a tree trunk cleft in two is fine. The holes that you drill into the mould will be on a curve, so a curved log can work well as long as it cleaves cleanly and will sit firmly, flat on the ground.

The idea is that it is heavy enough to hold the uprights steady, even when you are making a 6ft (180cm) high hurdle. In practice it is sometimes necessary to fix the hurdle down to the ground with ropes and pegs but this is only possible if you are working in an outdoor workshop and can peg down into the ground.

Mark out the positions of the holes on your mould, starting with the two end holes and make them 5ft 10in (176cm) apart. Mark a straight line between them with string. The centre hole will be 3½in (88mm) set back from this line. Mark the other holes evenly spaced (approximately 8½in/22mm), keeping a nice curve in your line as you go. The curve is necessary to create a flat hurdle. If you just made a hurdle on a straight mould, the tensions in the wood tend to make the hurdle 'wind' or twist at each corner. If you make the hurdle curved and flatten it, then it will stay flat. The books say things like 'leave your hurdles in a stack to flatten'. In our experience the novice hurdle-maker will take an awful long time to make enough hurdles to weigh each other down. Lay them out in ones or twos on a flat surface and put weights, such as logs or rocks, on for a couple of weeks and you will have lovely flat hurdles.

Use a ⅝in (16mm) drill and drill the holes right through the mould. They can be angled slightly forward by about 2½in (60mm) over a 4ft (120cm) high hurdle. This does make it slightly easier to weave a high hurdle and allows you to press the weave down with your knees (until it gets too high).

### Chopping Block

Your chopping block should be at a comfortable height for working. It is used to point the ends of your sales, to cut the ends of your splitting rods straight and to start the split going with the bill hook. A dome-topped block allows you to chop at a variety of angles

### Rod Horse

Somewhere sturdy to lean your rods upright is your next priority. You may be working in a shed, in which case just lean the rods up against the wall. The roof must be high enough to accommodate your longest rods or you will end up cutting them short, which is a shame. If you are setting up in the open, then a 'rod horse' is the answer. This can be just two forked posts with a sturdy pole suspended between them. It is a bit of an art to get a variety of length of rod to stand up steady against this pole. At a pinch you can lay the rods flat on the ground but remember this exercise is all about organizing yourself so that your material is instantly to hand without too much bending and walking about.

> **Tip**
>
> Always work on one side of the hurdle to minimize the amount of movement. You should stay on the split side of the hurdle, with the curve of the mould away from you like you are inside a bay window. All split sides and all cut ends are on this side, so the far side with the bark is the 'right side' when you come to put the hurdles up. This stems from the sheep-hurdle days when the rougher side was kept away from the sheep so as to not catch their wool.

### Splitting Post

The next on the list is a splitting post; this perhaps an optional item. I am sure that the very fastest hurdle-makers do not use a post to split on. If you can split a rod down with just a bill hook, brilliant. But a word of caution if you are new to hurdle-making: this is the most likely way that you will draw your first blood. It is hard to do without aiming the blade towards your unprotected hand. Far safer to start the split with a tap of a mallet on the back of the billhook twisting the hook from side to side to open up the split, then push the split rod onto the post. The post can be any handy wood driven firmly into the ground and shaped into a 'V'. A cleft oak log is ideal for this as it is already V- shaped and will be very durable. Just smooth the surface a little with a draw knife so that the rod will push through without snagging.

*Splitting post and chopping block all to hand.*
*(Photo: Anna Gray)*

The post now acts as a fixed point against which you can lever the rod back and forth, to keep the split running down the centre. If you are working inside, on a concrete floor, this can present a bit of a problem. One way round it is to set a square metal post-holder or a piece of cast-iron pipe into the concrete and shape the splitting post to fit snugly into this.

### Tip

Always split from the thin end to the thick as you are moving into more of the wood as you go. Remember that you must create a bend on the thicker side to bring the split back to the middle. If the split runs off too far just cut the rod square and start again. Perseverance is the name of the game.

## Hand-Tools

A bill hook is essential, a curved blade is useful if you are splitting the hazel down with the hook; for trimming, the best hook is small with a straight blade. Trimming up all your cut ends with a deft stroke of a sharp hook will definitely be the quickest way, but don't be too hard on yourself, a sharp pair of loppers is almost as good. You can cut your sales to length with your hook if you like but a hand bowsaw will be more accurate and create a clean square cut to get you started. A small handy mallet is great for starting the split (to tap the back of the bill hook) and a heftier one will help keep the weave battered down tightly. These do not need to be anything fancy but can be fashioned from the thick end of your hazel.

## Hurdle Pitch

Now arrange all your tools and devices so that you have the minimum amount of movement to get to them all.

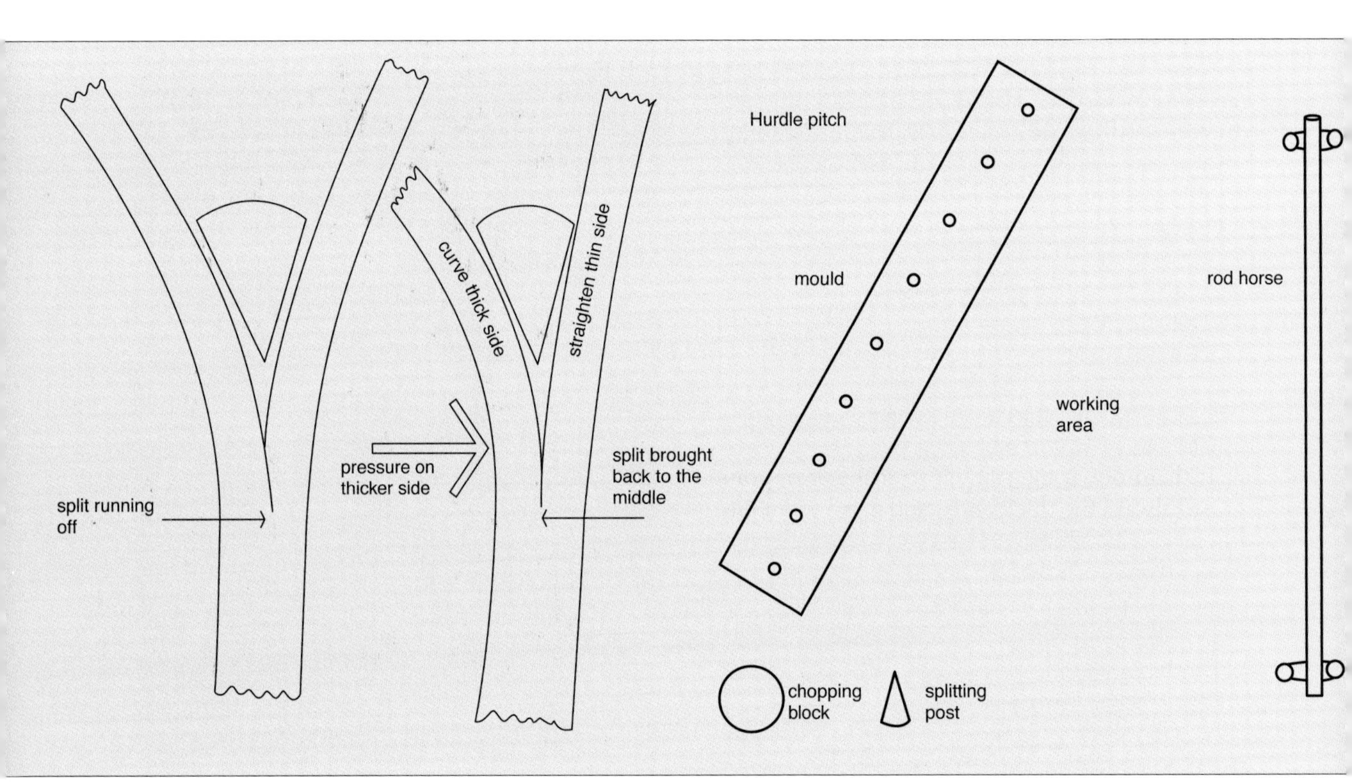


*How to keep the split running down the centre of the hazel.*

*Hurdle pitch set up so that everything is to hand.*

### Setting Up the Hurdle

You are now ready to begin!

The hazel should have been bundled in twenty or twenty-five rods and consist of a mix of sizes from very thin weavers to 2in (5cm) sales. Your first job is to untie the bundle and arrange your rods in ascending or descending size (by thickness not length). A bit of time spent sorting your rods now will save you lots of time later as each rod will be visible and easily picked up as it is needed. Your very thinnest rods will be used in the round for the top of your hurdle so can be set aside. The next section will be rods that are to be woven in the round at the base of the hurdle and they can be up to ⅝in (23mm) thick. Rods thicker than this will have to be split and your next section will have a range of splitting rods from ⅝in (23mm) to 1½in (40mm). There should now be just a few that are even thicker and these will be your split sales. Go back to the thinnest of your splitting rods and pick out two 1in (25mm) max rods as straight as possible to make the two round end sales.

First cut your sales to length. The length will depend on the height of your hurdle: a 4ft (120cm) high hurdle will need and extra 6in (15cm) for trimming. Do not be too ambitious when you are starting: a 3ft (90cm) hurdle is a lot easier than a 5ft (145cm) one. So whatever the height, add 6in (15cm) and cut all your sales to this same length.

Take the two 1in (25mm) diameter sales and, with the billhook point the thicker end, a nice tapering point about 3in (8cm) long. Drive these two sales into holes one and nine at either end of the mould, tapping them in firmly with the mallet. Then take the fatter sales and very carefully split them down the middle to make two, point these (always the thick ends) and drive them into the mould into holes two and eight. Repeat until you have filled all the holes (you will have one sale left over which you can save for your next hurdle)

You are now ready to start weaving the base of the hurdle. This is a bit complicated as you are tying these rods in, to make a firm base to your hurdle that will not drop out when the rods have dried, and you are carrying your hurdle around on the hillside.

Step one: Take two long rods from your thicker round weavers and place into the hurdle in the manner shown in the figure below. The tips of these rods form the tie rods on the right hand end of the hurdle. The butts will be cut off flush with the left-hand end.

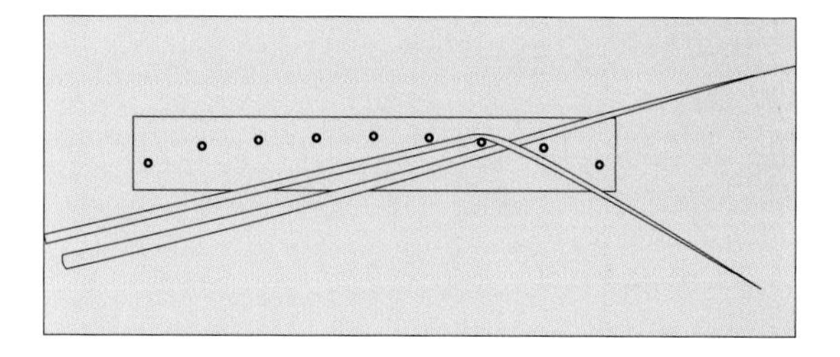

*The start of the weave.*

Step two: Working in the opposite direction, place the butt of the next weaver in front of sale six behind five and leave pointing out to the left. Repeat with three more rods starting each, one to the left of the previous rod (see figure above).

Step three: Tuck the last but one rod over and behind sale one and two; these two rods form the tie rods at the left-hand end. Take your loppers and trim all your butt ends close to the sales.

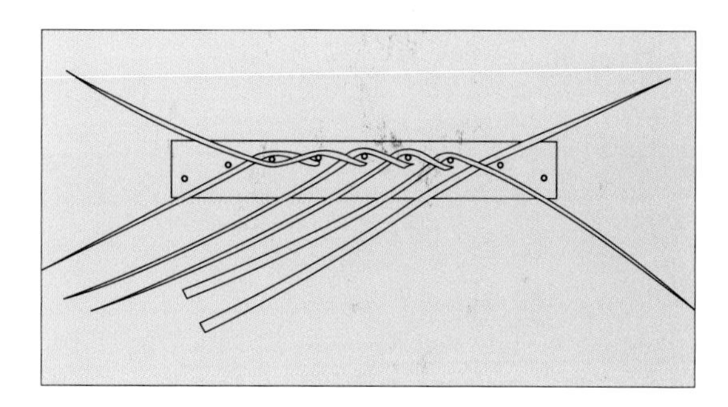

*Six rods in place ready to start.*

> **Tip**
>
> Keep your loppers to hand and trim ends as you go, cutting them with a sloping cut that leaves the rod neatly in line with the sale.

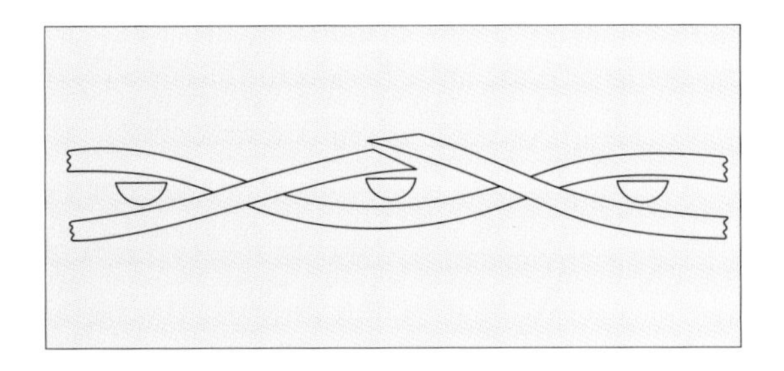

*How to trim the rods close to the sales.*

Step four: Ignoring the tie rods, pick up the first weaver from the left (it will be coming from in front of four going behind three), now weave it in front of two and behind one. Now the tricky bit – grasping the rod close to the sale, bend it against the sale until you hear a slight cracking sound. Straighten it out and, with two hands, twist the rod until the fibres separate and can be wound around like rope. Bend the rod right round and weave it back in and out, finishing up on the front of the hurdle somewhere in the middle. Repeat this manoeuvre with the next rod to the right. Weave the butt ends of the two right-hand tie rods back to the left-hand end. These will not be twisted round the end sale and woven back in but cut off flush. You will now have a triangle of weaving tapering down to the right hand end.

## Tip

Take a bit of time to get used to twisting the hazel rods. Working the rod back and forth a bit will loosen the fibres (tapping it with a hammer has the same effect). The trick is a good, strong grip – dry hands and rubbery work gloves can help, and a very slow and cautious approach until you are confident with the technique. Round rods should be turned through at least two 360-degree rotations. Split rods are more like 180 degrees.

Step five: You now fill in the triangle, working in the opposite direction. Take a new round weaver from your rod horse, place it with the butt in front of sale two and weave it in and out to the right-hand end, where you will twist it and return it to the centre of the hurdle. Repeat this with two more rods, each one starting one place to the right.

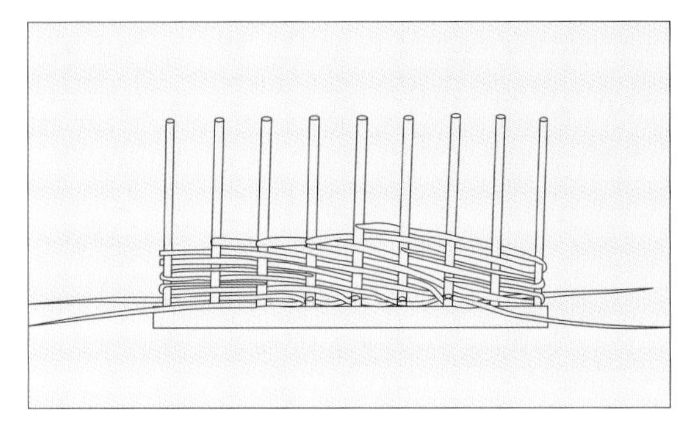

*Weaving back in the other direction.*

Step six: Pick up the tie rod nearest you at the right-hand end and pull it towards you to free it and then put it between sale eight and nine, twist it and weave it back into the hurdle. Repeat with the other tie rod, bringing it from the back of the hurdle from behind eight and in front of nine, twisting it and weaving it back.

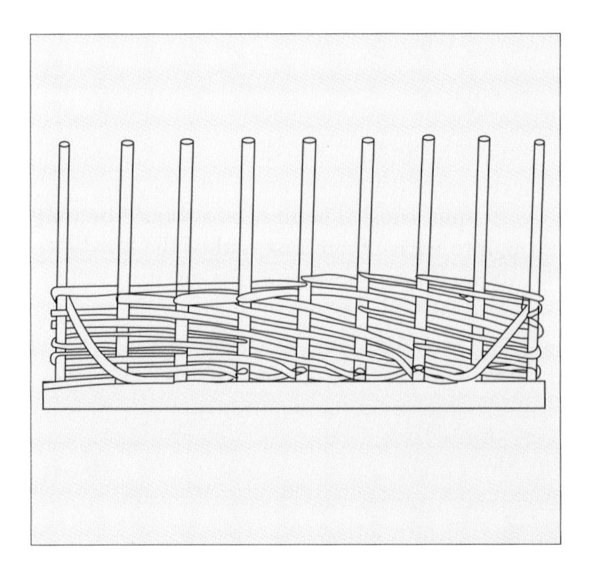

*The tie rods twisted and returned on either end.*

*The top weave half done.*

**Tip**

Start rods so that they alternate butts to the left then to the right to keep the weave growing level as you progress.

Repeat this procedure at the left-hand end with the two remaining tie rods. Ensure that all ends are trimmed neatly and compact the weaving down with a stamp of your foot. The weaving should be lovely and level.

Step seven: Now you are going to weave with split rods, so start with the thickest of the splitting rods (as long as they are not quite as thick as your split sales). The first two should be long enough to weave from one end of the hurdle to the other and be trimmed off at 6ft (180cm) Make sure that they are both the same way round (tips at the left hand end). The next split rod is also started with the butt right but this one needs to be longer so that you can twist it round sale one and return it to the middle of the hurdle. Trim close to the sale and start the tip end of its pair from that sale weave to the end, twist and return and with luck it will go all the way back to the left hand end. The next two rods can be 6ft (180cm) and be trimmed at the ends

Continue to weave the hurdle in this way, ensuring that the sales remain upright and are not pushed out of line by a thick weaver. If there are places where the sale is forced forward or backward (often where you have a join on a previous row), make sure that the next rod is sturdy and can force it back into line. Do not have joins in the same place in consecutive rows and do twist in the rods every third or at most fourth rod, to keep the edges of the hurdle robust. As the hurdle grows, the sales get slimmer but so do the weaving rods, so the weavers are always the ones that bend around the sales and not the other way round. When 5in (125mm) short of your finished height, finish weaving with split rods at the end of a complete row.

Step eight: You are now ready to weave the top of the hurdle. Start with the longer thin round weavers and, with the butt in front of sale eight, weave it from right to left and twist and return. Repeat with three more rods, each one starting one sale to the left. The last one has a double twist to form a knot and needs to weave back to the centre of the hurdle.

Step nine: This is the moment when, if necessary, you may move around to the back of the hurdle (if too high to reach over), push the butt of the next round weaver into the hurdle under two rows of weaving between sale two and three so that the butt rests against sale two; then weave it back to the right-hand end, twist and return. Repeat with the next rod one sale to the right (butt in front of sale three) twist and return. Place the third rod in front of sale four and weave to the end but leave it sticking out in front of sale eight and nine. The final rod is pushed in to sit in front of sale five and is woven to the end, twisted twice round sale nine to form a knot and then the tip of it is pushed in to the weave to sit in front of sale six and the rod is brought in front of sale eight, then the previous rod is also tucked back in behind sale nine.

Finally trim up any stray ends that have been missed and hammer the weave down tight. You can now trim the sales down to a level 1in (25mm) or so above the weave. Using the secateurs or a sharp knife, trim up any whiskery bits down the sides where you have twisted the rods and make sure the sides are all trimmed into a neat line. Lift the hurdle out of the mould by lifting the whole thing up at one end and knocking the mould off with your foot. Repeat at the other end until it is free. Now when you have admired it enough, lay the hurdle flat on a hard surface and weight it down to make sure it dries out flat. You can trim the pointed ends of the sales off, but this is not necessary.

### Continuous Hurdles

In some circumstances where hazel hurdles are required but square panels do not fit the space, you can put up a continuous hurdle, known also as an 'in situ' hurdle. Perhaps you want a fence on a slope and do not want the 'stepped effect' of panels or you have a curved section that needs a bespoke solution. This is where the continuous hurdle comes in.

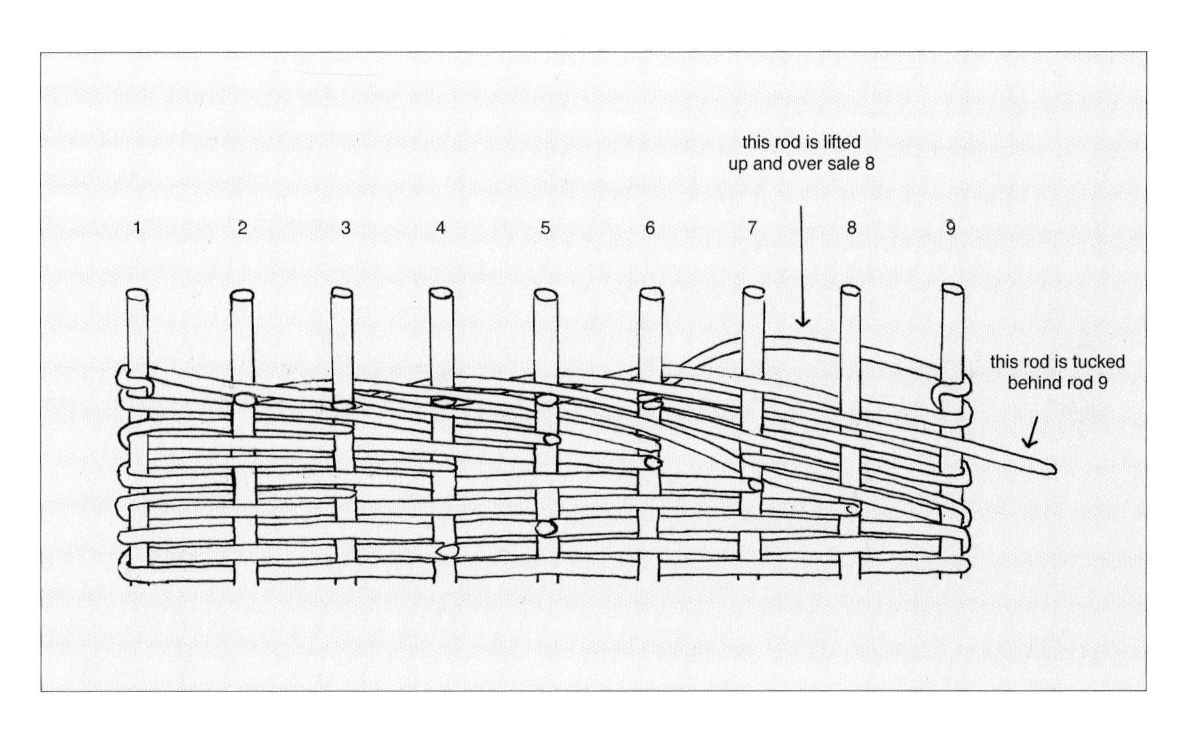


*Final stage of the weaving.*

ABOVE: *Hurdles being flattened.*
BELOW: *A continuous hurdle in a sloping garden.*

*Border hurdle woven on oak heartwood pegs.*

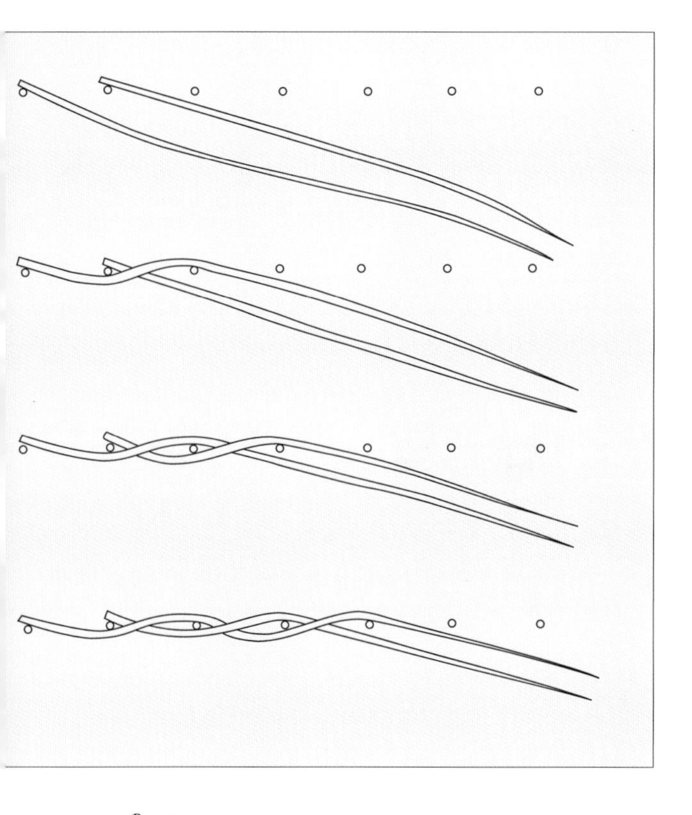

*Pairing.*

At its simplest you can hammer your hazel stakes or sales into the ground and just start weaving. The major disadvantage of this is that hazel does not last long in contact with the soil and the whole thing will no doubt blow down within a year or two. Far better is to make oak heartwood pegs about 2ft (60cm) long – sweet chestnut would be good as well, as both woods will resist decay in the earth. Cleave the pegs down with mallet and wedges or a froe and dress out with side axe and draw knife until you have a smooth peg with a diameter of 2in (5cm) maximum and a point on the end to go into the ground.

Mark out the line of your hurdle and make holes in the ground with an iron bar. The holes should be between 8in (20cm) and 12in (30cm) apart (10in/22cm is a good compromise). Drive the pegs in with a mallet so that they are one foot (30cm) in the ground and the same amount above.

Start weaving with round rods as fat as you can manage without pushing your sales out of line (1in/25mm max). The strongest weave is one called pairing where you take two equal-sized rods and weave them together in a spiral formation through the hurdle.

At the far ends it will be necessary to twist the rods and return them to tie the end sales in. You do not need to do this for every row, but every third or fourth. When you have reached the top of the oak pegs, take hazel rods and drive them down into the weave alongside each peg so that they will now take over as sales. Round rods of about 1in (25mm) are best for this. Now you can continue weaving with split rods just one at a time and taking care not to have running joints and to make sure that the weave builds evenly. Take into account the slope and build up areas that are to be higher than others at this stage. The top of the hurdle can be a repeat of the bottom section with round rods woven in pairs, and make sure that, in the final rows, the new rods introduced are tucked into previous rows so that they do not spring apart. Trim up all ends (do this as you go along and do not leave this all to the end) and cut the top of the sales cleanly with a saw. These fences can be reinforced by driving cleft oak posts or something similar into the ground along the line of the hurdle and wiring the hurdle to it.

*Finished hurdle with arch and gate.*

### *Westmorland Panels*

Known no doubt by other names in other regions, the Westmorland panel was championed by North-West coppice craftsman Colin Simpson. He decided that 4ft (120cm) wide panels were more practical if you were working with sub-standard hazel and did not have the lovely long rods needed for the traditional panel. The top and bottom rails are drilled to take the sales and are therefore integral to the finished panel. Because the sales are held in place by these rails there is no need to wind the rods around the end sales to bind the panel together. The trick is to build the panel with the top rail off, so that you can push the weavers down from the top but keep the sails in line, then you can fit the top rail back on at the end. You need to keep trying the top rail on, at regular intervals to make sure that it will still fit.

### *Plant Stands*

The best way to make wigwam structures for the garden is to use a template or 'mould' to insert the upright hazel rods – vehicle tyres are ideal for this purpose. Pick a tyre size to suit the diameter of the finished structure. A tractor tyre that measures 4ft (120cm) across can be used for large stands 7ft (210cm) high, standard car tyres

*Westmorland panels.*

*Hazel plant stands.*

for 6ft (180cm) stands. The smallest stand practical to make with hazel is about 5ft (150cm) high and 14in (35cm) across, but these take a lot of patience and careful selection of hazel to get rods that are flexible enough to weave so tightly.

Whichever size of mould you choose, you will have to start by drilling holes in order to push your uprights in and hold them firmly, but not too tightly, or it is very tricky to get them out again! Car tyres contain metal, so use a drill bit that you are not too attached to – size around 1in (20–25mm). An electric drill will make this job a lot easier. Mark the tyre with evenly spaced holes, normally an odd number: seven for the smaller stands and nine for the large one.

Select straight hazel rods for your uprights – they must not be too thick for your drilled holes but need to be sturdy enough not to be bent out of shape by the weaving, ¾–1in (20–30mm). Bring them together at the top and tie firmly with string. Select only the thinnest, long whippy rods for weaving. Hazel that is in rotation is either too young and brittle or too thick, so a quiet walk through an old stand of derelict hazel is best where you can spot the occasional sun shoot that has grown up from an old stool and struggled up and up, growing only slowly widthways, so remaining very thin and pliable. These rods are characterized by a lovely silvery colour, which suggests slow growth and plenty of annual rings for strength. Take only one or two from each stool, as these rods are important for the regeneration of the hazel.

*Nine straight hazel rods held in a tractor tyre for making a plant stand.*

**Weaving**

You will need a bundle of twenty weaving rods, spread the rods out and sort them into pairs. They need to be of equal thickness and length, if possible. Take a pair of long weavers and place them either side of your first upright.

You are aiming to get a double twist into the hazel pair, which means that each time the same rod goes behind each upright and likewise the other travels around the outside. Work anti-clockwise and do two rows that are pressed down tight against the tyre. It will be necessary to join in a new pair of rods as the first ones run out. To do this, aim to overlap by a couple of sections, weaving the old and the new rod together as one. Try not to have new rods joining in at the same place each time you weave round the tyre but stagger the joins to avoid weak points. On the third time around, lift the weaving up gradually to create a rising spiral, keeping a steady 10in (25cm) from the previous row.

As you get towards the top of the stand it will get tighter and tighter to weave and will challenge your hazel weavers to bend without breaking. Use the finest, most flexible rods you can find. At the top, join the pair of weavers in with the seven uprights and bind the nine rods with a wire tie. Besom wires or the long version of the bag ties that farmers use for sealing paper potato sacks are best for this and can be twisted with a ratchet action tool.

When held firmly in place, take a fine hazel rod and start twisting it to break the fibres and create a withy or binder that can be wrapped around on top of the wire three times and tucked tightly down to hold the end in place. Now take a pair of secateurs or loppers and trim up all the ends. Carefully pull the stand out of the tyre mould and stand back and admire.

*The double twist so that one rod is always on the inside.*

*Bag pull tool and wires.*

*Hazel plant stand in author's garden.*

*Hazel plant stands in a border to support sweet peas.*

*Coracles based on the Boyne currach.*

## *Coracles*

The Irish Boyne currach was always made from round rods, most often hazel but sometimes willow. A renowned basket weaver, Olivia Elton-Barrett, took a transcript of the last Boyne coracle maker Michael O'Brien and recreated the Boyne currach in a slightly scaled down form shown in the picture overleaf. The coracles in the picture are for one person and are a similar size to the coracles you find in Wales. They are covered in calico and painted with bitumastic paint. The true 'Boyne' was 6ft (180cm) long, used by two people and it was covered in a large cow hide.

### To Make a Boyne-Type Coracle

Take twenty-eight hazel rods, maximum diameter ¾in (2cm) and push them into flat, level ground (lawn turf is best for this), so that they form a rounded oblong 4ft by 3ft (120 × 90cm). Eight rods along each of the long sides and six along each of the short sides. Weave thinner rods in pairs all the way round for one row, then cut a plank to fit across the middle, taking a small chunk out of each corner to make a tight fit between the rods. Carry on weaving in pairs over the seat and round until you have at least three rows to secure the seat in place. Then bend the rods down to make the bottom of the boat, tying them to the seat to hold them in place; the overall depth of the boat should be 14in (42cm).

Tie each rod to its corresponding pair, keeping the shape of the bottom of the boat as flat as possible. When you have bent and tied all eight pairs on the long sides, loosen your ties again and, trimming the tip end to length, insert it down into the weave on the far side of the boat. Repeat this process with the twelve rods on the short ends of the boat, bending, tying and tucking until all rods are securely held in place.

Lash each crossing point using square lashing and sisal string for preference, as it grips as you tighten the knot.

Work from the centre out and keep the crossing rods square and evenly spaced. When all

*Paddling your coracle.*

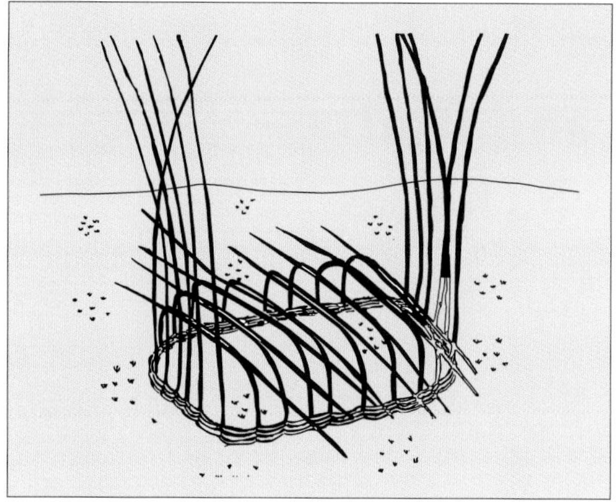

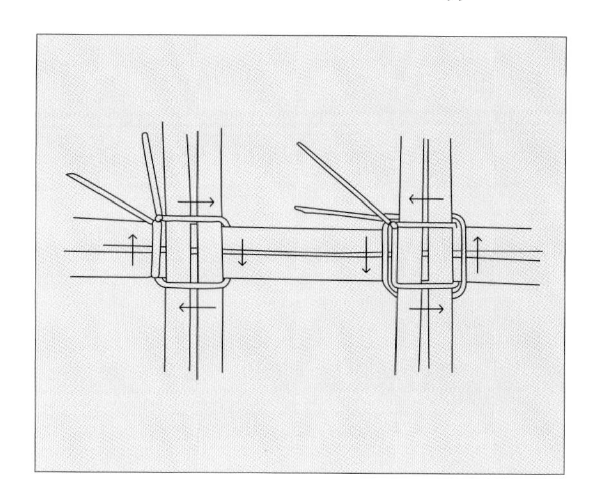

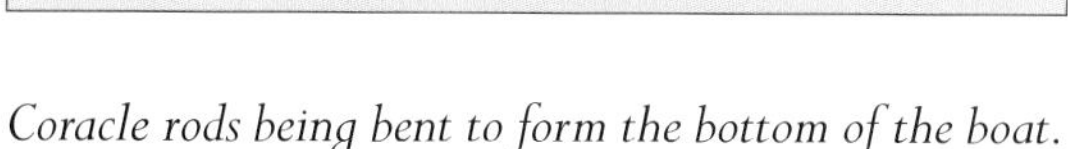

*Coracle rods being bent to form the bottom of the boat.*

*Square lashing.*

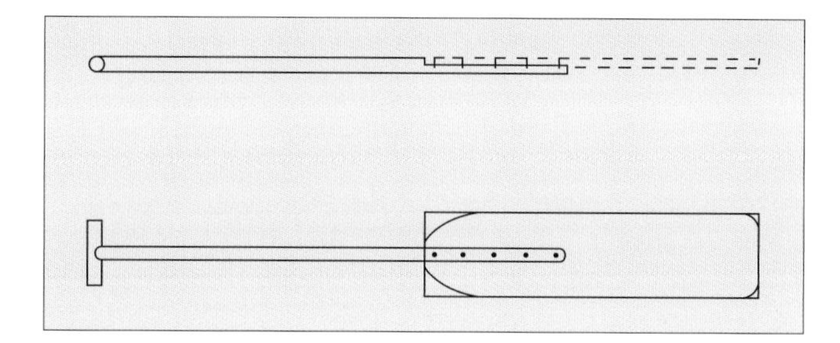

*Coracle paddle.*

forty-eight crossing points are lashed securely, take an iron bar and lever the boat up out of the ground (you may wish you had not pushed them in so far at this point). The hazel will want to expand at this point so you need to tie the sides of the boat together under the seat to keep it safely supported by the woven rim. Trim all the ends close to the weave to create a smooth top to the boat leaving a half inch or less protruding.

Cover your boat with a sheet of calico 2½ yards (225cm) of 72in (183cm) wide material and sew it in place with strong thread, folding and tucking the fabric to make it fit smoothly.

Two coats of bitumastic paint will make it waterproof – a fun but messy job.

Traditionally a paddle would be carved out of a cleft piece of ash but a simple paddle can be made from exterior plywood and a broom handle. Take one side off the broom handle with a draw knife to fit the plywood paddle flush and cut a piece off the other end to make a cross bar for good grip.

The trickiest part of coracle making is learning to paddle it. You must scull it with a subtle figure-of-eight motion over the front of the boat, that draws the water towards you and makes the boat move forward.

## CONCLUSION

Coppiced hazel is such a wonderfully versatile material to work with, whether it is a bundle of whippy rods for weaving into a strong fence or a boat or a collection of simple walking stick blanks with a beautiful variety of bark colour from silver grey to red brown with a delightful range of mottled colours in between. A bundle of smooth stakes for hedge laying or straw-bale building, or maybe a spray of perfect fan-shaped tops for weaving into natural supports for your herbaceous border. Nothing in nature betters hazel for any of these jobs.

# Chapter 7
# Coppice Woods and Crafts – Variations on a Theme

## ASH WOODS

Ash (*Fraxinus excelsior*) is a true native of Britain and is found in almost all regions. It grows best, however, on rich, moist but well-drained soils over chalk or limestone. But, as Edlin points out in his wonderful Woodland Crafts in Britain (Edlin 1973), it is these base-rich soils that are also prized for agriculture and as a consequence ash woods for timber production are a fairly rare sight. The Morecambe Bay area is characterized by its carboniferous limestone, which stands out in rocky outcrops with thin or absent soil. Ash trees grow here but are often slow-growing and stunted where their root systems are restricted by the rock. In between the rock formations, where the soil collects and has some depth, ash will grow freely and fast with wide growth rings; qualities that are preferred for the ash crafts.

Ash will coppice well; however, the new shoots that grow from the cut stool can have a bendy (sometimes contorted) growth form, which is fine if you are looking for a specific curve for a chair but less useful for yurt poles or straight handles. Ash does grow from seed very freely and if what you need is straight, small-diameter poles, a thicket of ash saplings can be the best place to look. There are estates in the South of England that grew pure ash coppice cut on a fairly short rotation for hop poles and crate rods (Edlin 1973); in the North it would have been cut for crates too and also bobbin wood (wooden cotton reels).

*Ash stand on limestone.*

*Ash gate hurdle. (Photo: Anna Gray)*

### *Properties of Ash*

Ash has long been prized for the qualities of its timber. It does not have a long life in contact with the earth and, therefore, it is not usually used for fencing, except for temporary stakes for hedge laying. But it is great for just about everything else: turning, cleaving, planking, building, furniture. Tool handles are one of the uses for which ash is most favoured. It has strength and flexibility, so is able to absorb the shock received by an axe or a hammer. Wooden tent pegs are often made of ash for the same reason and are still in demand for marquees, as they grip the soil better than a metal peg. Ash is in demand these days for yurt roof poles and even the crown ring – in fact any craft where steaming is required to shape the wood. Having extolled its virtues for all the crafts it feels bad to mention this but, of course, it does make the very best firewood: its low water content at felling means you can burn it green, and seasoned it is even better. Charcoal too is an option, but that really is a sacrilege!

### *Coppicing Ash*

Ash will often be favoured for standards within the coppice. So before you take your chainsaw or bowsaw out, have a good look and select the best ash for leaving – not just the mature trees but perhaps groups of saplings that will grow on to make standards in the future. One of the benefits of ash standards, and the same could be said of oak too, is that they come into leaf late and so the woodland plants that are adapted to completing their flowering early (bluebells, anemones, wild garlic, etc.) have maximum light when they need it. Likewise coppice underwood species, such as hazel, get a head start too.

> The oak before the ash, we will only have a splash,
> Ash before the oak we are bound to have a soak.

Ash that is to be coppiced should not be cut to ground level, as the new shoots or regrowth will

*Ash pollard, completely hollowed out by age.*

form under the bark of the base of the stem; so leave 2–3in (5–8cm) of trunk standing as a stool – no more than that or the stools as it ages will get unstable and unable to support the regrowth. If the stool has been coppiced before, then it is advisable to leave 2–3in (5–8cm) above where it was previously cut, thereby avoiding cutting back into the old timber at the bole. There is always a risk cutting back into very old wood, where the bark is too thick and/or there are fewer viable buds to form new shoots (Harmer and Howe 2003) – this is why old coppice stools can be so large.

One of the specialities of Cumbria is the old ash pollards that can be found on field boundaries. These trees can be of great antiquity but have been regenerated repeatedly over many centuries by cutting the branches back to a bole that is well above the head height of the stock, either sheep or cattle. That way a 'crop' of nutritious leafy branches could be cut and fed to the animals in a controlled way on a regular basis.

**Ash products**

| Product | Number in bundle | Ideal dimensions | Price range |
|---|---|---|---|
| Hedge stakes | 10 | 5ft long (1.5m) 1–2in (2.5–5cm). | £3.50–£5 |
| Beanpoles | 10 | 7ft (2m) long 1–1½in (2.5–4cm). | £5–£7 |
| Besom handles | 20 | 3ft 6in (1m) long, up to 1in thick and smaller for children's brooms. | £10 + |
| Strawbale pins | 10 | 4ft long (1.2m) up to 1½in wide (to 4cm). | £6–£8 |
| Yurt poles (roof), peeled or unpeeled | 10 | Up to 8ft long (240cm) and 1¼–2in diameter (3–5cm). | £10–£20 |
| Rake and broom handles | Individual | 6–8ft long (180–240cm) | £1–£2 each |
| Walking sticks | Individual | ½–1½in (2–4cm) up to 6ft (180cm) long. | £1 + |
| Axe and hammer handles | Log | Up to 4ft long (120cm) 6in (15cm) diameter. | £2-£5 |
| Furniture | Log | As required. | £2 + |
| Sports equipment | Log | As required. | Various |
| Gate hurdles | Log | 4in (10cm) minimum 6ft (180cm) long. | £2 + |
| Tent pegs | Log | 4in (10cm) diameter, 6in to 2ft (15–60cm). | 50p–£2 |
| Hay rakes | Handle poles and log to cleave | As required. | Various |
| Wood turning | Log | As required. | Various |

### *Hedgestakes, Bale Pins, Beanpoles*

Dress them out with a bill hook to remove side branches, cut to length and bundle up in tens or sometimes eleven for beanpoles, so that you have one extra for placing along the row.

Occasionally you might want an odd, long length for a pole-lathe pole or a special building project, so do not cut them to length until you are sure what they are for.

### *Yurt Poles*

Yurts are a popular structure for a temporary (or sometimes more permanent) shelter in a wood, as they blend into the surroundings in a pleasing manner. Originally used by nomadic people over an extensive range from Turkey through to China, it is predominately in Mongolia (where it is known as a Ger) that these portable tents are still in use. The walls are made of a lattice work of coppiced poles and the roof poles are sturdier ash poles that are sometimes steamed to give the yurt its characteristic domed shape. Instructions here focus on producing the raw materials for a yurt from coppice poles.

Yurt roof poles should be a maximum diameter of 2in (5cm) and minimum of 1in (2.5cm) if you are supplying someone else, then they will normally undertake the peeling and bending. However, if you are making your own yurt, you will need to consult a more detailed instruction manual, but the basic preparation of the poles is

*A yurt makes an ideal temporary shelter in the woods.*

as follows. Remove the bark and smooth the pole with a draw knife. If felled while the sap is rising the bark should peel quite easily leaving a smooth rod. Working on the thicker end, flatten the profile of the rod on one side until the rod is just 1in (2.5cm) thick, this should extend for 2ft (60cm) from the end. The rod is ready now for steaming.

The simplest steamer is made from a plastic drainage pipe with the ends blocked off and a flexible pipe attached. The other end of the flexible pipe is attached to a vessel for boiling water and a heat source applied to create the steam. The poles are placed inside the pipe and all steam is sealed inside as best you can. They will require steaming for about forty minutes.

You will need to prepare a jig to shape the poles on and this can be done with some large logs fixed firmly to the ground with pegs and ropes. There is an enormous amount of pressure created from the bent wood, so do make sure that your jig is robust enough. The top end will need tying down to a sturdy pole that is suspended off the ground, again strong enough to with stand the forces of the steamed wood.

Work quickly when you remove the steamed poles, trapping the thinned end under the first log and bending it over the second log. Tie the top end down to your crossbar, lashing it firmly with strong cord. Try to ensure each pole has the same profile as you work. Steam and bend a small batch at a time and leave to set in the jig for a couple of days.

The thinner end is shaped to fit into the crown wheel, usually with a square profile to keep the poles from twisting around.

The wall rods are also bent to shape, though they do not normally need steaming. Peel and smooth rods with a maximum diameter of 1in (2.5cm). Ash is usually too chunky for this, so you may want to use hazel or willow. Then leave them to set curved in a jig that is just three poles driven firmly into the ground in a straight line and the wall poles threaded between them in alternating direction.

These rods are drilled with equally spaced holes and then threaded together with cord to form a trellis that can be collapsed down to a bundle of poles for transportation and quickly pulled out to form the walls.

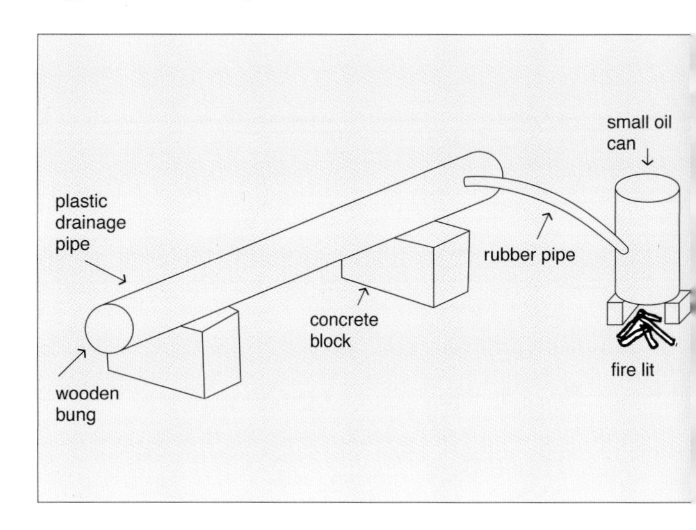


*A simple steamer.*

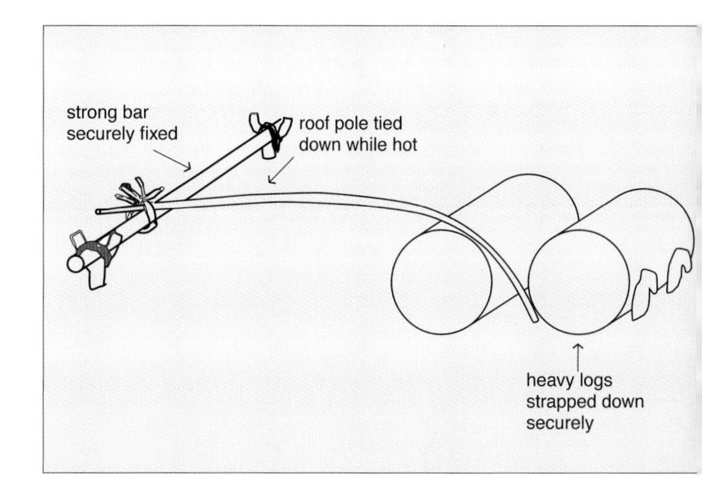


*Yurt roof poles being set in shape.*

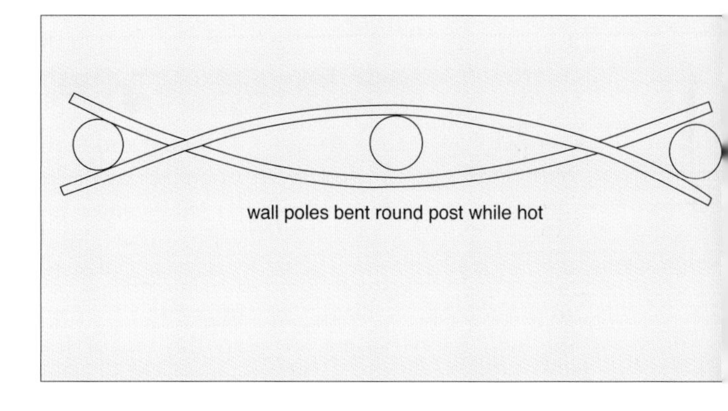


*Yurt wall poles being bent to shape.*

## *Cleft Ash*

Coppiced poles may be cleft to break down larger diameter wood into usable billets for different products, such as tent pegs or tool handles. Beware though that when a tree has grown four or five trunks off one stool, the centre point of the wood will be offset because the tree will grow faster on the side facing out. This is not necessarily a problem; however, for some jobs, a maiden tree that has been able to put on even growth rings may be the best option

### Hay Rakes

Traditional hay rakes are designed to be light and well-balanced, a perfect ergonomical design for a day's toil in the hay field. Hay rakes are constructed from four components:

1. The handle or stail. This can be a round pole, either ash or hazel, with the bark on or peeled as you see fit. It needs to be straight and smooth. You can start with a pole that is bent and, when the time is exactly right, you can straighten it by securing the end in a brake and applying pressure to the curved section (too soon and it will spring back – too late and it might snap). The optimum time for doing this is about six to ten weeks from the time it was cut. One way of smoothing a handle down is to use a stail engine, which is sprung loaded so that it will expand as the handle tapers from thin end to thick. The stail engine works a bit like a giant pencil sharpener and shaves the bark off and smoothes the handle as it turns.

2. The head. An ash log is cleft down into slim segments and then shaped with a drawknife until you have an oblong bar 3ft long (90cm), 1in (2.5cm) wide and ¾in(20mm) thick. This will need to be seasoned before the rake is assembled.

3. The brace or bool. One style of rake has a steamed hazel rod 3ft long (90cm) and ¾in (20mm) diameter that is steamed to make it supple and bent into a lovely U shape on a jig shaped for the purpose and tied until it is set. Steaming the hazel can be done on a small scale using a length of plastic pipe and a kettle. The pipe will need sealing to keep the steam in. Refer to the Yurt section of this chapter for more details about steaming.

4. The teeth. These are also ash wood about 4in long (10cm) and cleft down into regular squared shapes. A good tip for this is to cut your ash log to length and mark the top of it with lines ½in (12mm) apart in both directions. Then bind the log round with string and start splitting along the lines using a froe and a mallet. The binding holds the log together while you split it in both directions so at the end you have a bundle of peg blanks. Discard the outside pieces with bark on.

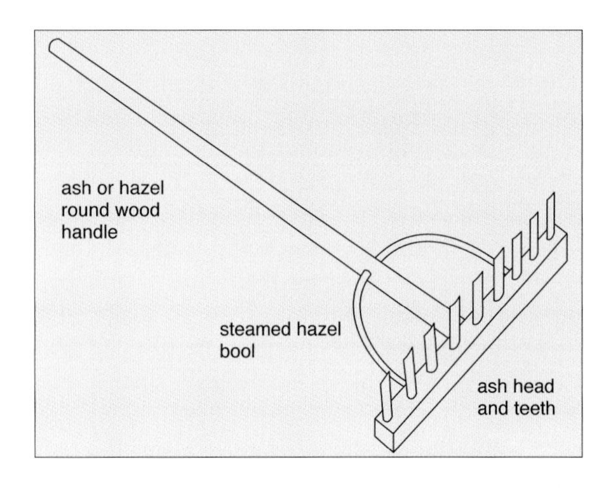


*A Northern style hay rake with a bent hazel bool.*

The square blanks are then hammered through a tine cutter, which can be as simple as a piece of pipe that is hard enough to file an edge on it. This will cut the soft green ash wood and turn the square section into a cylinder. The rake has fourteen teeth; leave them to dry out and shrink.

To assemble, the handle must be fitted to the cross bar. Drill a hole in the crossbar with a 1in-bit (25mm) and whittle the end of the handle down until it fits snugly into the hole. Drill the handle with a ½in (12mm) bit, 10in (25cm) from the end and thread the hazel brace through. Mark the position of the ends of the brace where they meet the crossbar, drill holes and shave the brace down to fit tightly into the holes. Finally, drill the crossbar with regularly spaced ⅖in (10mm) holes and fit the teeth. A spot of wood glue will help keep the teeth in place. Traditionally all the joints were drilled through and pinned with cleft ash pins.

*A tine cutter for making the ash pegs for hay-rake teeth.*

## Tent Pegs

Cut your ash log to peg length. This can be from 6in (15cm) to 2ft (60cm), and cleave the log using metal wedges and a wooden mallet, splitting it in half, quarters and eighths, if the log is big enough. You aim to end up with a triangle of wood pretty much peg sized, 2in (5cm) deep, 1in (2.5cm) wide for a standard 12in (30cm) peg.

Now shape your peg – this should take eighteen strokes! This is only possible if you have a stock knife; if you are using a bill hook or small axe it will definitely take more.

First take the bark off the back of the peg (two strokes), then shape the point that will go in the ground (five strokes), then shape the nose and chamfer the top (six strokes), then take a saw and cut into the peg at an angle to make the beak no more than a third of the way through. Then cut away the excess and trim the sharp edges where the rope might rub (five strokes).

This will give you a sturdy peg that should not split when hit with a mallet.

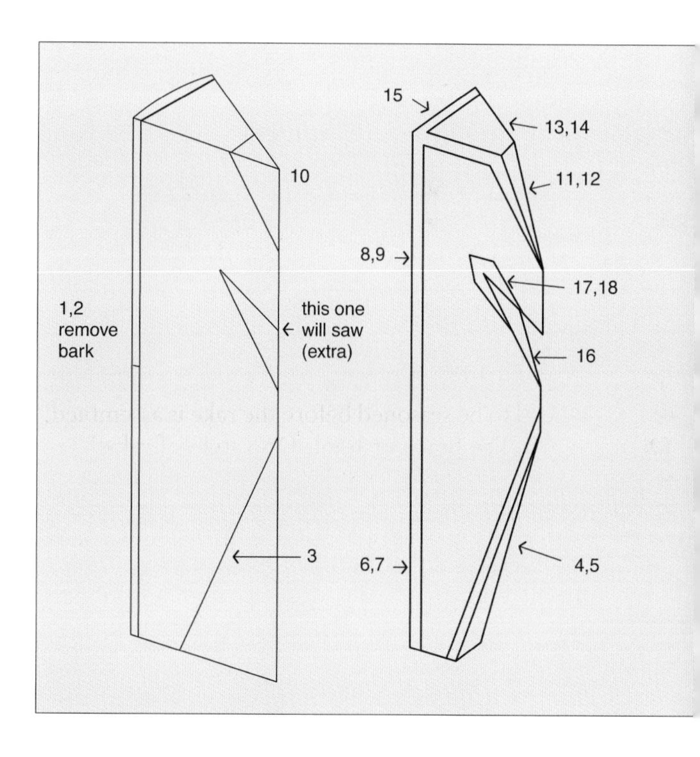


*Tent peg making – 18 cuts.*

*Tent peg making with a stock knife.*

## OAK

Oaks are the quintessential trees of Britain, and are even considered the national tree of Britain, by some. When thinking of oak, one has a mental picture perhaps of a tree that has grown in open parkland with spreading branches and a broad but imposing profile. In fact it is surprisingly common to find oak growing in dense woodland of multiple-stemmed stools that speak of a coppicing history for this versatile tree. There are two main species of oak native to Britain: the pedunculate oak or English oak (*Quercus robur*) and the sessile oak (*Quercus petraea*). English oak is the one with the best timber qualities and has been favoured as a standard tree in coppices over much of Britain. Sessile oak is the predominant species in oak woods on the upland, acid soils of the North and West, but in fact the two species mingle throughout woodlands up and down the British Isles.

### *Properties of Oak*

Oak is our most durable native timber, but even so it is only the hard heartwood that will withstand rotting when in contact with soil – the softer sapwood should be removed. The growth form of oak varies from the clean, straight grain of a fast-growing maiden tree, perfect for cleaving, to a more contorted, twisted growth of some coppice regeneration that has its own charm for use as rustic poles for creative structures.

### *Coppicing Oak*

Oak will coppice freely when young, though the regrowth is prone to mildew and can become rather distorted. Older oaks can succumb to 'stump mortality' (see Chapter 4), so be aware of this if restoring coppice that has not been cut for forty or more years. As with most broadleaves it is important to leave a short stool for the buds to break from, so cut at a height of 2–3in (5–8cm) above the ground or above the height of the last cut. Oak is a favoured food for deer and in a mixed coppice will be eaten in preference to most other species.

### *Coppiced Oak Products*

Historically, oak woods were coppiced to make charcoal for iron smelting, to produce bark for tanning leather and in a few areas, notably the Furness district of Cumbria and in Worcestershire, to make swill baskets or spelk baskets from riven oak. All these products are still relevant, if only on a fraction of the scale that they once were produced. These days though the peeled poles that are a by-product of bark peeling are more likely to be turned into rustic creations for a ready garden market.

**Oak products**

| Product | Sold as | Ideal dimensions | Price range |
|---|---|---|---|
| Peeled oak | Long poles | 1–6in diameter (2.5–15cm). | 50–100p per foot |
| Oak bark | Bundles | Various. | 7–10p a kg |
| Gate hurdles | Logs for cleaving | As required. | £2–£5 |
| Rustic furniture | Peeled poles | As required. | 50p + per foot |
| Furniture | Logs for cleaving | As required. | Individually priced |
| Swill or spelk baskets | Logs for cleaving | 6in (15cm) 4–6ft long (120–180cm). | £1–£5 |
| Pegs for continuous hurdles | Logs for cleaving | 2–4ft (60–120cm). | £1–£5 |
| Cleft oak fencing | Logs for cleaving | As required. | £1 per ft + |
| Fence posts /gate posts | Logs for cleaving | As required. | Individually priced |
| Shingle or shakes | Logs for cleaving | 12–18in (15–20 cm). | Individually priced |
| Lathes for building (restoration) | Logs for cleaving | As required. | Individually priced |
| Timber framing | Whole timbers | As required. | Individually priced |
| Wood turning | Logs for cleaving | As required. | Individually priced |

## *Peeled Oak for Bark and Rustic Furniture*

In the spring when the sap starts to rise, usually about mid-April, though the date creeps earlier with climate change, it is possible to lift the oak bark off the trunk as easily as peeling an orange. Coppiced oak is favoured because the bark of young trunks cut on a 20–25-year cycle contains the most tannin. There is a tannery in Devon that uses oak bark to make high-quality leather and once every other year, a lorry comes up to the North West to collect the bark that has been peeled and stored.

The peeled coppice poles have a market too for rustic furniture and garden structures.

### Tools

A peeling iron is used to remove the bark; it is like a blunt chisel with a slight curve and a

*The bark man Brian Churchill.*

*Peeling oak for bark and rustic furniture.* (Photo: Mike Carswell)

strong handle to grip, and it is possible to buy one from a tool supplier. Edlin mentions the use of bone to make a peeling tool, a horse's tibia was filed down to make the chisel and was thought to be the very best tool for the job. A more modern improvization is a piece of copper pipe hammered flat and then shaped.

## Method

Score the trunk along its length using the peeling iron and then, utilizing the curve of the tool, work the bark loose all the way around. If the sap is flowing well, the bark will lift away without any trouble in large sheets that curl up as they dry. Sometimes the sap is not flowing well and the bark feels sticky and it is a great effort to remove it, in small unsatisfactory scraps. This may be because the temperatures are too low – even in mid-peeling season, a cold spell can slow the flow of sap. It may be because the tree you have felled for peeling has been suppressed by its larger neighbours; unless the tree has a vigorous canopy of leaves you will not get the sap movement you need for peeling. This is one reason why coppiced oak is favoured where the regrowth is all of an age and receiving maximum light. It is pointless to struggle with a sticky tree, try a couple more and if they are not peeling, leave it until the next fine day.

## *Peeled Poles*

Oak used to be peeled with the tree still standing, up to the height a person could reach, prior to the tree being felled. Unlike cork oak (Quercus suber), where the bark can be harvested and the tree lives on, peeling the bark on the English oak will soon kill it. The normal practice is to fell the trunks and, if they are not too heavy, place them on an improvised bench, which lifts them up to a comfortable working height. Leave the trunks as long as possible if you are to use the poles for rustic products and do not dress out all the side branches but peel them with forks intact, as these can be incorporated into a design. Peel right down to an inch diameter, if it is peeling well. The tradition is to sell the poles by the running foot, so some system of measuring the finished lengths is important to put in place as you go. It is a good idea to leave them standing upright stacked against a tree to allow even drying.

### *Bundling the Bark*

Collect the bark up using the largest sheets to enclose the smaller pieces and tie securely with string. If you have been fortunate and the lengths of bark are long and large, it may be necessary to roughly cut to length with a bill hook so that your bundles are a manageable size. The bark should be stored under cover out of the rain, otherwise all the tannins will be washed out and the bark is useless for tanning.

### *Uses of Peeled Oak*

The peeled poles can be used for a variety of rustic constructions; it can be very effective if you utilize the natural forks and shapes of the wood in your design. It is a good idea to remember that young poles are mainly sap wood and they will not have as long a life as outdoor furniture if not protected with a wood preservative.

*Detail of using peeling iron (copper pipe).*
*(Photo: Mike Carswell)*

*Oak bark bundled.*

### Rustic Arch

Take four peeled oak posts 9ft long (275cm) and of similar diameter to each other. Lay them out on a flat surface in two pairs 18in apart. Cut cross pieces from peeled poles that are somewhat more slender and can be interesting shapes; cut them to 2ft long (60cm), so that when they are attached to the upright posts they protrude 3in (8cm) either side. They are attached by drilling a guide hole and nailing. Alternatively, drill and screw but it will look best if you counter-sink the screws in ⅜in (10mm) holes, at least ½in (1cm) deep and then plug the hole with a whittled round dowel in a hard wood like oak or yew. Squeeze a drop of wood glue into the hole, drive the peg in tightly and trim flush. When you have your two side sections made to match, you will need to secure them in position by digging out holes and sinking the uprights into the ground. Backfill with rocks and tamp the earth tightly around the posts. Bridge the gap overhead either with a simple straight section constructed like the sides or choose some lovely matching curved sections to create an arch. Secure with screws and plug the countersunk holes with a wooden peg.

## SWEET CHESTNUT

Chestnut is undoubtedly an important coppice tree in Britain. Whereas stands of chestnut can be found as far north as Scotland, regeneration from seed is infrequent in northern parts. This may well change if global trends of climate warming continue. As a coppice crop it is found mainly in Kent, Surry and Sussex, where it has been grown intensively as simple coppice for hundreds of years. Introduced by the Romans, who prized the nuts for food, it later became widely grown in those southern counties, as the timber proved very useful too (Rackham 1986). Chestnut, unlike hazel, does not like the chalk soils of the South Downs but favours more acid soils and thrives on the sandstone uplands of the High Weald.

### *Properties of Chestnut*

Associated with the hop gardens of Kent, chestnut poles were grown to support the hop vines,

*Rustic oak pergola. (Photo: Mike Carswell)*

as the wood has exceptional longevity in the ground. It grows fast from coppice stools if the conditions are right and the wood is mainly heartwood with very little sap. This heartwood surpasses oak for its resistance to rotting when set into the ground.

### *Coppicing Chestnut*

Chestnut will produce a ready crop of coppice poles but, as with all the species dealt with in this chapter, the longer the coppice has been neglected, the more risk there is of the coppice failing. Up to thirty or forty years there should be no problem, after that stools begin to be less reliable and produce weaker regrowth. Happily, chestnut can be regenerated through layering, just like hazel, and this is a good way to fill gaps caused by stump mortality. It will also reproduce from seed, so that is another way of increasing your stool density. If you have a neglected chestnut coppice, be bold and bring it back into rotation, as the dense spacing and monoculture of a coppiced wood will not develop into attractive high forest.

### *Chestnut Coppice Products*

Hops were grown until the end of the nineteenth century, up slender poles 14ft long (4.25m) and only 1in (25mm) diameter at the top. Each plant needed two poles, so that meant 2,000 per acre, which would have had to be regularly replaced (Edlin 1973). From 1900 onwards the need for thin poles was replaced by a more modest requirement for larger poles to support the network of wires and sisal string that the hops grow up. These in turn have been replaced by tanalized softwood.

Happily the chestnut industry has been able to adapt. The second most important property is its ability to cleave cleanly – hence the rise of the cleft chestnut paling fence, which dates from 1905 and is still in production today. Cleft from poles cut on a 10–14-year cycle, up to 25,000 pales can be produced from 1 acre of chestnut, which is enough for 1 mile of chestnut fencing (Edlin 1973). The market for cleft chestnut fencing has been kept alive largely by the construction industry, which has long favoured it for temporary fencing, as it is durable, semi-self-supporting, light and portable, and can be re-used. Small exclosures of chestnut fencing can be quite effective against deer browsing in coppiced areas. The deer will walk around a circular fence and, though they can jump over a standard 4ft (120cm) height, they are unlikely to as it appears to be a trap. The beauty of it is that it can then be moved to a new area once the coppice is above deer-browse height, though this is harder than it sounds, as the bramble grows through it and holds it fast. One major market is the armed services, as it has been used to create track ways for tanks on manoeuvre in the desert.

### Chestnut Paling

Chestnut is cut to length in the wood, depending on the height of the palings required and then taken back to a yard to cleave. The poles are cleft down into segments using a froe or 'dillaxe', as it is sometimes known, and a mallet with some leverage applied using a cleaving break to keep the split running true.

The pales are then trimmed using a drawknife and shavehorse to take the bark off and to make them even. Then a blunt point is

*Chestnut coppice.*

**Chestnut products**

| Product | Sold as | Ideal dimensions | Price range |
|---|---|---|---|
| Hedge stakes | 10 | 5ft long 1–2in (2.5–5cm). | £3.50–£5 |
| Beanpoles | 10 | 7ft long 1–1½in (2.5–4cm). | £5–£7 |
| Paling fences | Logs for cleaving | 2–6ft long (60–180cm). | |
| Peasticks | 20 | 5ft long (150cm) | £5+ |
| Rustic poles | Poles with bark or without | | |
| Yurt poles (roof) peeled or unpeeled | 10 | Up to 8ft long (240cm) and 1¼–2in diameter (3–5cm) . | £10–£20 |
| Rake and broom handles | Individual | 6–8ft long (180–240cm). | £1–£2 each |
| Walking sticks | Individual | Various. | Various |
| Building | Timber/poles or cleft | Various. | Various |
| Furniture | Cleft logs | As required. | Various |
| Trugs | Cleft and steamed | As required. | Various |
| Gate hurdles | Cleft logs | As required. | Various |
| Fence posts and gate posts | | | |
| Post and rail fences | | | |

made with an axe on the top end of the pale. Finally the pales are bundled tightly and left to straighten and season.

The wiring of the paling is an operation that is largely mechanized today but can still be done on a small scale using a 'long walk' similar to that used for making rope. It needs to keep the wires taught at the correct spacing whilst the palings are fed by hand in between the wires. A trolley with a crank that can be turned to twist the wires and grip the paling tight is wheeled along the length of the wire.

## MIXED COPPICE

In most native woodland there is a mixture of species, the particular composition of which is affected by a number of factors:

- Soil type and pH and the underlying bedrock.
- Ancient woodland or plantation.
- Management favouring certain species.

As all our native species, and most of the more contentious late arrivals, are suitable for coppicing, a mixed wood presents no barrier to reinstating a coppice regime. It does, however, present a bit of a headache. If the wood is mixed with less than 50 per cent hazel, you will need to make a decision on whether it should be cut on a short hazel coppice cycle (5–9 years), where you will end up with a lot of thin poles of oak, ash, sycamore, birch, hornbeam, etc. that might make low-grade hedgestakes but not much else; or whether it should be cut on a 10–15-year cycle, which will make the hazel overstood and less valuable, but might provide other products, such as birch besom for horse jumps and brooms or yurt poles, or maybe bobbin wood if you happen to have a bobbin mill in the vicinity. Perhaps it would make more sense to go for a 15–20-year cycle, when at least it will be producing some decent poles for fencing or rustic work and furniture, and the less useful produce can be charcoaled or even used as firewood. If firewood is your main product, you require optimum size poles of 5–10in (12–25cm) diameter, in which case a cycle of 20–30 years is the most appropriate, especially if there is a very small percentage of hazel.

It is possible to run two cycles in one coppice area, for instance oak and hazel coppice. Cut your hazel at 7 years and your oak every third cut at 21 years. You will inevitably lose some of the vigour of the regrowth on your hazel as the oak stools start to dominate the canopy and reduce the light levels, you will need to monitor the growth rates to maintain the health of your coppice. In a similar way you might coppice hazel at 7 years and birch at 14 to maximize the productivity of your wood.

### *Common Trees of Mixed Coppice*

#### Sycamore

Often maligned as non-native and invasive, sycamore is actually a very useful coppice tree. The vigour of the regrowth on a sycamore stool is very impressive, getting away above the deer browse height in one year (deer do not seem to like it much if there are tastier options available). However, it can then succumb to the ravages of the grey squirrel, something to be aware of if you are trying to grow a good crop of sycamore poles. Though the poles are fast growing, they do not have much strength or longevity. Once they start to mature into poles of 15+ years, they are great for cleaving and turning, especially for novices, as the wood is still quite soft and sappy. As timber it can be beautiful; keep a sharp lookout for ripple effect in the wood of any standards that you fell, as that is very sought after by furniture makers. Of course it is brilliant for charcoal and firewood.

#### Birch

Birch is another tree with a bad press, especially from a commercial forestry perspective, where it has been seen as a weed and ruthlessly eradicated from plantations. An early 'ruderal' species, it has masses of wind-born seeds. Birch will colonize open ground, such as new plantations where the soil has been disturbed, or newly coppiced woods, especially when the open phase has been extended by browsing, suppressing the regrowth of stools. This is not necessarily a problem as the young thickets of birch can be cut on a hazel cycle of 7–9 years and still produce good besom for horse jumps or brooms. They are probably at their optimum though on a cycle of 10–20 years, especially if a bit of selective thinning has gone on to allow the crowns to develop into a good shape.

Whilst birch is young, the branches have a lovely upward pointing sweep, and each piece is usable: the longer ones from the bottom can be sold for the jumps at racecourses and the prime branches at the top dressed out into bundles for besom brooms, all a perfect yard (1m) long. The remaining poles can be used as beanpoles, or used for charcoal or firewood, so there is no

*Cleaving chestnut palings with a froe or dillaxe.*

*Fifteen-year-old sycamore coppice.* (Photo: Anna Gray)

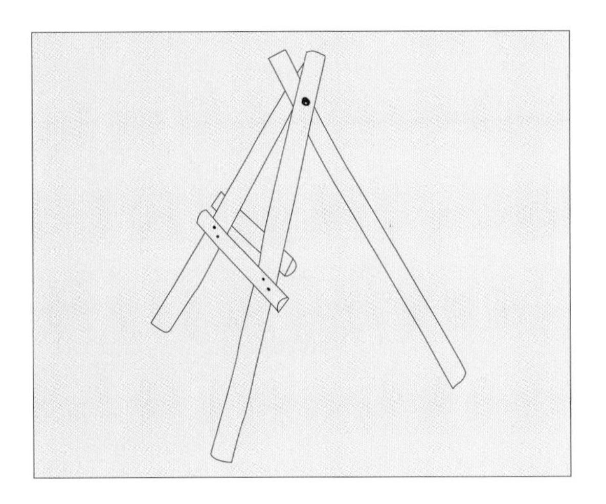

*'A' frame cleaving break, portable and strong.*

*Birch trees – young on the left, mature on the right.*

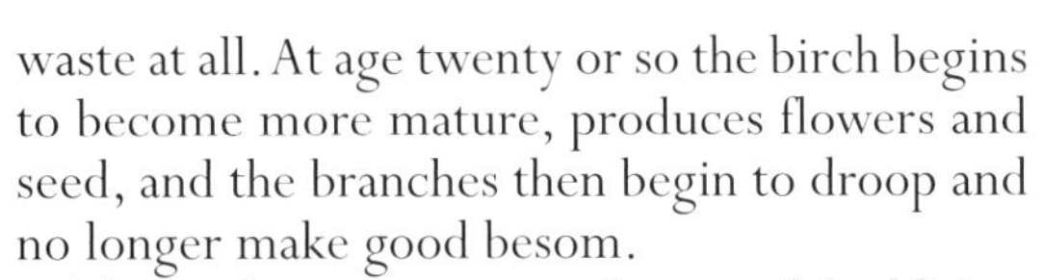

waste at all. At age twenty or so the birch begins to become more mature, produces flowers and seed, and the branches then begin to droop and no longer make good besom.

The timber continues to be a useful addition to the charcoal and firewood piles. Perhaps because it does not grow quite so straight here, we seem to be less inclined to use birch in Britain in the many ways that they do in Scandinavia – the timber for furniture and the bark for containers and clothing and just about anything you can imagine. But do have a go at being creative with birch and remember to tap the sap for wine in early spring too.

All this has little to do with coppicing apart from the fact that up to about thirty or so years the cut stool of birch should regenerate. However, because its primary method of regeneration is by seed, more mature trees can fail to coppice. Even on younger stools, the regrowth can be weak and of poor form, but usually there are so many young seedlings that this is not really a problem.

### Yew

Yew is, of course, an evergreen tree but should not to be confused with a conifer, most of which do not coppice. Not perhaps a common tree of coppice but where it occurs on limestone, it is abundant and does coppice well. The poles that spring up are potentially fast grown and knot free and can be used for long-bows or rustic work to great effect. If you have some yews within your mixed coppice you might be tempted to leave them as standards but occasionally there are just too many, so do not be shy of coppicing them but consider taking up long-bow making to utilize the poles that are produced.

### Beech

Beech, like birch, has a great capacity for seed production and is reluctant to coppice when mature; even when younger than forty years of age, it may well have weak regrowth and high stool mortality (Harmer and Howe 2003). The beech woods of the Chilterns, which are famed for their tradition of 'bodging' or chair leg production, were managed on a selection system with replacement trees being brought on from natural seeded regeneration (Edlin 1973).

So if you have beech mixed in with other species in your wood, should you coppice? You may decide to leave mature trees as standards but do remember that they are shading trees that have very little undergrowth associated with them, so when they are mature no shrub layer will survive. They will dominate a swathe of your coppice without much benefit, apart of course from aesthetics because they are undeniably an attractive species. Mature trees will produce copious seed in good mast years and, if you are in part of the country that is not considered native territory for beech (everywhere except south of a line from Norfolk to south Wales and Devon), then encouraging this species to spread in ancient semi-natural woodlands is often frowned upon.

## Alder

Alder (*Alnus glutinosa*) is most likely to be found growing in groups, in a damp patch within a wood or by the side of water courses. Wet alder woodland is known as 'carr' and is most likely to be growing as a grove of multi-stemmed trees. This may well be because it has been coppiced, but even left to its own devices alder is one tree that just cannot help itself coppice. Old stems are often ringed around with new growth just waiting for the mature tree to fall and leave canopy space for the young shoots to fill. One big advantage of alder coppice is that it is largely untouched by deer browsing, unless they are really hungry. So sometimes it is possible to get away with no fencing. Traditionally used for underwater structures like piers or riverbank reinforcement (revetments), it is for clog soles that alder is famous. Soft and easily carved when green and light and strong when dry

Not particularly favoured for firewood, alder makes exceptionally good charcoal. Light but hot burning and at one time essential to the manufacture of gunpowder.

## Willow

Willow within woodland is most likely to be goat willow (*Salix caprea*); it can grow to quite a large tree. It has merit both aesthetically (the flowers 'pussy willow' are one of the thrills of spring), and from a conservation point of view, as there are many species that live and depend upon it. As timber it is less useful, disparaged as firewood and charcoal (it is soft and breaks up when you bag it), and hopeless for fencing as it will grow roots wherever it is allowed contact with the soil. Historically, of course, people were always resourceful, and trugs, barrel hoops, tool handles and gate hurdles were all made from willow at one time (Miles 1999).

It would be churlish to have a book on coppice without a mention of the osiers, or the basket willows. Salix viminalis and Salix triandra to name just two, are grown on an industrial scale for the willow-basket trade and more recently for biomass. If you have a damp corner and wish to grow some of these species, and there are many variations of colour and form, then the main small-scale use will be for 'living willow' structures. Utilizing the readiness for willow to strike as cuttings, rods are cut in the spring and pushed directly into the soil where their long regrowth can be woven into fences, houses, mazes, tunnels, as inventive as your imagination will allow. Living willow structures are a very popular activity to carry out with children in school grounds and parks. The best aspect from a business viewpoint, is the ongoing maintenance, as it will continue to grow at a rate of about 10ft (3m) a year and these shoots will have to be woven in or cut back to maintain the sculptural quality of the work. A job for life!

*Birch bark is a very versatile material.*

The third category of willow that deserves a mention is white willow and crack willow (Salix alba and Salix fragilis). Large trees of riversides and meadows, they were in the past pollarded to produce a mass of boughs for fodder and a variety of farm uses from hurdles to mangers. Cricket bats are one willow product that is still current and the variety Salix alba var. calva is known as the cricket bat willow.

## Coppice products from mixed coppice

| Product | Sold as | Ideal dimensions | Price range |
|---|---|---|---|
| Hedge stakes<br>Sycamore, birch, etc. | Bundle of 10 | 5ft long 1–2in (2.5–5cm). | £3.50–£5 |
| Beanpoles<br>Sycamore, birch, etc. | Bundle of 10 | 7ft long 1–1½in (2.5–4cm). | £5–£7 |
| Besom handles<br>Sycamore, birch, etc. | Bundle of 20 | 3ft 6in long (1m), up to 1in (2.5cm) thick and smaller for children's brooms. | £10 |
| Rake and broom handles<br>Sycamore, birch, etc. | Individual | 6–8ft long (180–240cm). | £1–£2 each |
| Walking sticks<br>Blackthorn, holly, etc. | Individual | 1in (25cm) by 4ft (120cm) long. | Individually priced |
| Furniture<br>Any | Log for cleaving | As required. | Individually priced |
| Gate hurdles<br>Sycamore, elm | Log for cleaving | As required. | £1 per ft + |
| Wood turning<br>Any or various | Log for cleaving | As required. | Individually priced |
| Besom brooms<br>Birch | Bundle | 3ft long (1m). | £5+ |
| Besom for horse jumps<br>Birch | Bundle 12in diameter (30cm) | 5ft long (150cm). | £3–£5 |
| Long bows<br>Yew | Poles | 4–6in (10–15cm). | £1 per ft + |
| Rustic<br>Yew | Poles | 1–5in (2.5–12cm). | 50 pence + per ft |
| Bark or bast for chair seating or rope<br>Elm or lime | Log | 6in (15cm) diameter. | £1 per yard/m |
| Clogs<br>Alder | Log for cleaving | 8in (20cm). | £3–£5 |

*Yew coppice.*

*Colin Simpson carving clogs soles with a stock knife.*

**Besom Brooms**

Take your bundles of prime birch besom that you have cut from the crowns of the birch tree and had seasoning somewhere dry but with a good air circulation for at least 6 months. The sprays should be a yard long (just under a metre) and still retain some of the delicate curve that they had when growing. Start to build up your broom, placing the besom around the core of the broom with the curve facing inwards. You aim to build a broom that is tulip shape rather than a bunch of flowers.

As the bundle gets too big to hold in your hand, cradle it in the crook of your arm like a baby and continue to add birch, turning as you go to maintain the lovely shape. When you judge that you have enough, place the bundle in a besom clamp and close the clamp to squeeze the besom tight. The clamp should be roughly half way from either end of the broom head. Turn and tease the besom to improve the shape and add a little more if required. Holding the bundle compressed with the clamp, thread a wire around the bundle up close to the clamp. A 10in (25cm) wire will make a good size besom, but if you want a large besom, then use a 12in (30cm) wire. Twist the wire tight with a 'bag pull tool'. Repeat this process twice more, moving the besom so that the next wire is about 4in (10cm) towards the base of the broom head and the final one the same. Now take a bowsaw and very carefully cut the waste ends off about 1in (25mm) below the final wire. Your broom head is about done, a little trimming of the top ends is allowed but do not overdo this as you want to avoid a stubby end. Just gather together the tips and chop off any excess with a bill hook on a chopping block.

For a handle you will need a coppice pole, nice and straight and about 1in (2cm) in diameter. Hazel is ideal but ash or even young birch can be used. Bark can be left on or shaved off according to your taste. Cut to 3–4ft long (90–120cm), depending on the height of the besom user, and put a long tapering point on the thinner end. Now drive this point into the centre of the broom head and tap it down into place, taking care that it is running down the middle and not showing on one side. Tap it down firmly until you judge that the point is in beyond the wire that is around the middle of the broom. Finally trim any stray bits that are sticking out and stack your finished brooms standing upright on their handles or bundled together with a few more to maintain their shape. Besom brooms are perhaps a little out of favour. There are not

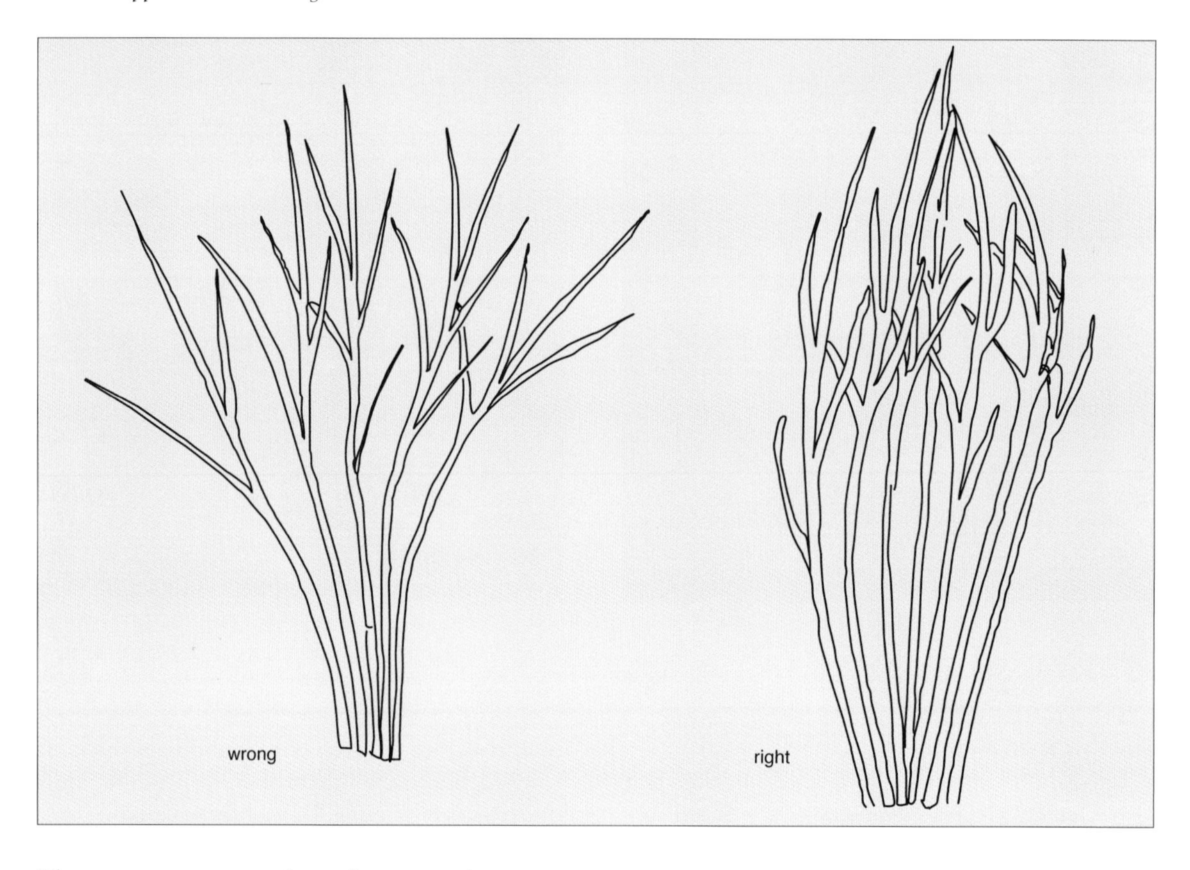


*The correct way to put a besom broom together.*

many parks departments that bulk-buy them these days but they remain the very best broom for sweeping up outside as they are equally good at spreading worm casts on your lawn or lifting damp leaves from your paths.

For a more aesthetic broom try binding the birch with brambles stripped of their barbs or strips of ash or willow wood cleft and pared down to a fine lath.

## Birch for Horse Jumps

Whilst you are preparing besom for brooms, it is ideal if you can also make bundles for horse jumps. These utilize the slightly coarser, longer branches from lower down the tree. They need to be 5ft long (150cm) and 12in (30cm) diameter where they are bound with string. These bundles are used in great quantities by race courses and if you have a ready supply of young birch and can get a contract, then it is a valuable addition to a coppice business.

## Elm Bark Seating

Sadly elm trees are a more rare sight in our woods since the scourge of Dutch elm disease. The English elm (*Ulmus procera*) has been particularly badly affected. Wych elm (*Ulmus glabra*) has a more northern range and has, to some extent, shown resilience to the disease. It coppices well and this is a valid way to support elm trees in combating Dutch elm disease, as the beetles (*Scolytus scolytus*) that are responsible for spreading the fungus (*Cerotostomella ulmi*) fly at a height of about 20ft (6m), therefore the coppice regrowth remains quite healthy until it reaches that height. In addition, the bark has to reach maturity before it can host the beetle, therefore keeping the elm young by regular cutting will also help.

If you want to harvest elm bark, however, you do need a tree that is at least 6in (15cm) diameter. Fell it and sned it in the spring when the sap is rising, leaving the trunk as long as possible.

*Chair made by Paul Girling with elm bark seating.*

The outer bark must be removed with a drawknife leaving the inner 'bast' intact. Then take a sharp knife and score the bast down in long, evenly spaced strips of about 1in wide (2.5cm). Now peel the bast back and it should come off cleanly in long strips that can then be coiled and hung up to dry. You can use it for seating straight away or use it dry; however, you will then have to soak it in hot water to make it usable. The bark of small-leaved lime (*Tillia cordata*) can be harvested in the same way.

## CONCLUSION

It is perhaps a tall order to condense all the variety of coppice trees into one chapter, with such a wide range of potential products and material for crafts that these trees can provide. We have attempted to give you a glimpse of the range of possibilities that coppice wood presents. Innovation is the watchword of a modern coppice business and, whilst honouring the skills and techniques of the traditional crafts, one should always seek out new and creative ways of using coppice materials products.

# Chapter 8
# Wood as a Fuel

## FIREWOOD

### *Introduction*

The act of lighting a fire holds a fascination for most people. It is such a primeval task and so intricately bound up in our basic survival instinct. We have sadly become removed from this pleasure, with our flick-of-a-switch electricity and gas or oil central heating on tap. Motivated by the fear of fossil fuels running out and the inevitable rise in fuel prices, more of us are turning back to wood for heating. But can we meet this increase in demand for firewood? Do we have enough woodland to supply logs to all these newly installed woodburners? The answer I believe is a guarded yes, but only if we bring more of our woods back into active management. In this chapter we will look at how you can produce your own top-quality firewood from leafy tree to glowing ember.

*Woodburning stove.*

### Which Wood to Burn?

You can burn as fuel both hardwoods and softwoods. Hardwoods have a denser structure, whereas the open structure of softwoods means that there are more air spaces in the wood. The amount of heat that you get from a log is governed by the amount of moisture in that wood; the fact is that if you oven dry wood and have a moisture content (MC) of 0 per cent then you will get just over 5kWh (kilowatt hours) per kilogram (kg) of wood. This is true whatever the species; the thing to watch out for though is that a kilogram of hardwood is a much smaller volume than a kilogram of softwood with all its air spaces within. So you end up burning a large quantity of softwood logs to get the same heat output. The term for the amount of heat a fuel will produce is calorific value. The calorific value of green wood (unseasoned or 50 per cent MC) is 0 and the calorific value of oven dry wood is 5; the firewood you produce will be somewhere between those two values.

The other quality of softwood that makes it less favoured for firewood is the tendency for it to spit when burning. This is important if you are burning in an open grate but less trouble in an enclosed stove. Some hardwood species are also liable to spit, including sweet chestnut.

Logs to burn; logs to burn;
Logs to save the coal a turn.

Here's a word to make you wise
when you hear the woodman's cries;
Never heed his usual tale
That he's splendid logs for sale
But read these lines and really learn
The proper kind of logs to burn.

Oak logs will warm you well,
If they're old and dry.
Larch logs of pinewoods smell
But the sparks will fly.
Beech logs for Christmas time;
Yew logs heat well;
'Scotch' logs it is a crime
For anyone to sell.
Birch logs will burn too fast;
Chestnut scarce at all;
Hawthorn logs are good to last
If cut in the fall.
Holly logs will burn like wax,
You should burn them green;
Elm logs like smouldering flax,
No flame to be seen.
Pear logs and apple logs,
They will scent your room;
Cherry logs across the dogs
Smell like flowers in bloom,
But ash logs all smooth and grey
Burn them green or old,
Buy up all that come your way
They're worth their weight in gold.

Honor Goodhart (1926)

There is a lot of good advice in the old poem, including the reminder to buy up ash logs as you can 'burn them green or old'. This is because the moisture content of the green ash logs is already very low.

## *Coppicing for Firewood*

Coppicing for firewood is a perfectly sustainable woodland-management system. The rotations are usually between 15 and 30 years. The issues are the same as with all the coppice variations that we have considered:

- Species.
- Light levels.
- Stocking density.
- Deer browsing.

### Species

Most of our native hardwoods are suitable for coppice management. If you have access to an area of conifers, you will have to consider thinning or clear-felling and restocking. In a mixed hardwood stand, then the only barrier to reinstating a coppice regime will be if the trees are too old and likely to be moribund (see Chapter 4). Ideally you will have in your firewood coupe a good proportion of ash, but sycamore, birch, hawthorn, cherry and hazel are all useful species. Oak and beech make good firewood but they take longer to season and you may be in a hurry.

### Light Levels

This is a question of the correct standard density. Coppice requires good light to grow, so make sure you have reduced your standards to allow vigorous coppice regrowth. There is detailed discussion of this in Chapter 4.

### Stocking Density

If the trees are sparsely scattered, don't worry too much as there will be plenty of seed available from nearby trees, which will immediately respond to the increased light levels and germinate to fill the gaps with seedlings. If what you end up with is a thicket of birch regeneration, then you may wish to augment the mix with some judicious clearing and planting.

### Browsing

It could be argued that fencing of firewood coppice is not as important as if you were trying to produce perfect coppice rods or poles. A bit of munching and forking will not matter if it is

*Mature firewood coppice.*

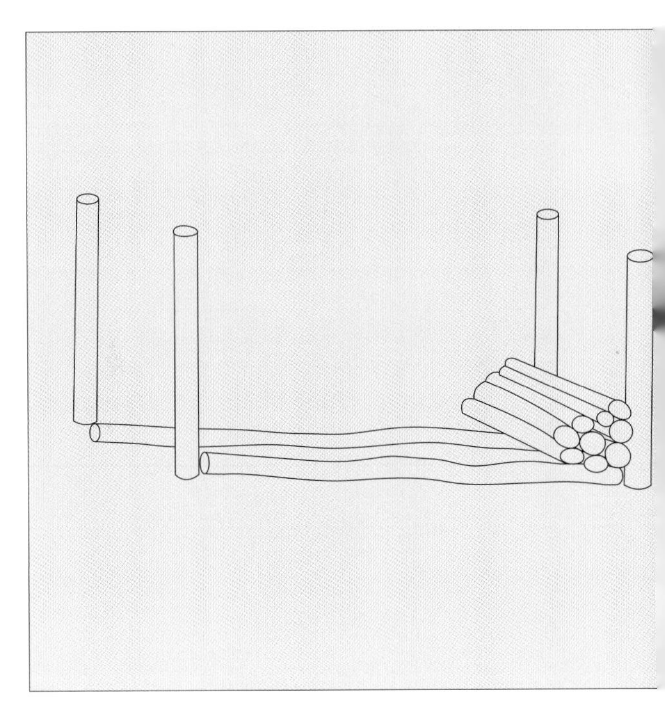

*Stacking rack in the wood.*

going to be cut as firewood. But do not let yourself be lulled into thinking that you can ignore the deer, as any length of time where the new shoots are being browsed back, will weaken the overall regrowth and require an extension of the rotation length. In any case, straight stems are easier to process by hand or machine.

### Felling the Coupe

Trees are usually felled in winter when the leaves are off the trees; the moisture content is already lower if the sap is not flowing and that gives you a head start with the seasoning. Also, you will avoid disturbing nesting birds, as detailed in Chapter 4. So the cutting season will be October to the end of February. How much of the tree you use will depend on your own time/labour/effort equation. Ideally you will use it all, with the main trunk being cut to length, all the side branches cut off and cut to length down to about 5cm (2in), then the rest bundled up to dry out for kindling. If you are embarking on a more commercial enterprise, then it may be more practical to process all the wood bigger than 8cm (3in) and stack, burn, chip or scatter the brash to be rotted down back into the soil.

### Seasoning Firewood

As mentioned earlier, wood has a high percentage of water in it, expressed as moisture content (MC). In order for it to burn efficiently, you must allow this moisture to leave the wood, a process known as seasoning. The length of time required depends on the species of the wood and the size of the logs, but for firewood, a minimum is one year (not forgetting that ash will burn 'green or old'). One-year seasoning or 'summer dried' is generally plenty for species such as birch, sycamore and hazel. Harder woods, such as oak, beech, elm and yew, will require at least two years. It is tempting to say the longer the better but do bear in mind that wood starts to break down and go soft and spongy when it is too seasoned; birch in particular can be past its best by the end of two years, as naturally occurring fungi feed on the sugars in the wood and reduce the calorific value of the timber.

*Timber at ride side waiting to be moved.*

The timber is first stacked in long lengths in the wood and, if it is to remain there any length of time, it is worth constructing a simple stacking device.

The traditional woodland stack was known as a cord. Firstly, two posts are driven into the ground to support the stack at one end. Logs are placed along the ground lengthways to keep the stack up off the ground. The logs are cut to 4ft (1.2m) lengths and placed into the rack to a height of 4ft (1.2m); each running length of 8ft (2.5m) represents a cord. Another way to measure wood is by a rough guide to the weight. The stacks can be constructed in the same way as a cord but by having the logs 6ft 6in (2m) long and the stack 3ft (1m) high, each running meter of the stack is about one tonne. The rule of thumb is 2m3 of stacked wood equals one tonne. This must be a rough measure, as the wood is drying out and getting lighter all the time and the actual weight cannot be accurately measured this way. The aim at this stage is to allow free airflow throughout the stack. It does not matter at all if it gets constantly dowsed with rain, as the wetting and drying seems to speed up the seasoning. The timber may remain in the woods like this for up to two years, but bear in mind that many woodland creatures will take up residence in your piles and be disturbed or destroyed when you do move them. A better alternative would be to remove the timber to a yard or nearer the house and keep it in similar stacks until you are ready to process it.

## Processing Firewood

Seasoning will be speeded up if you cut your logs to length and chop them. The cutting can be done with a saw horse and a bowsaw. Cut the logs to the optimum length for your stove. Or, if you are unsure, a standard log is about 8in (22cm) long. Some small grates or pot-bellied stoves will require the logs to be cut even smaller. A bowsaw will cross cut green logs with ease as long as it is sharp. The blades do require replacing very regularly unless you are using one of the older types that can be sharpened.

Using a chainsaw for logging will definitely speed things up but please undertake the basic training for cross-cutting before you begin

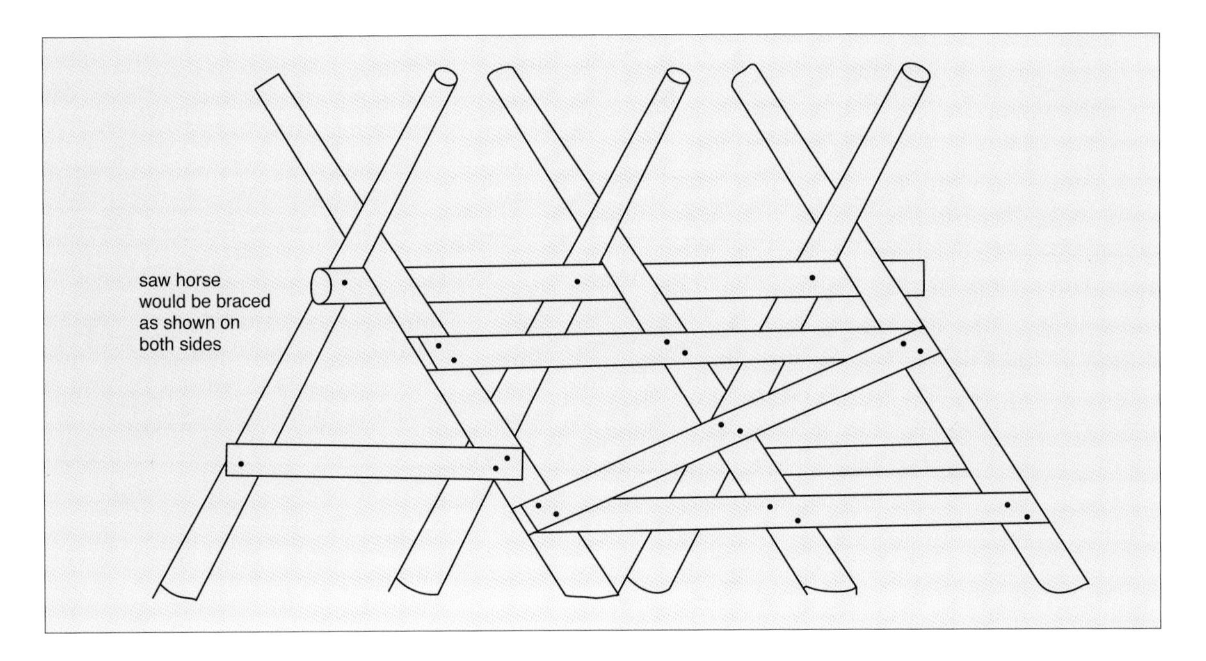


*Sawing horse.*

(CS30). Having the correct protective clothing (trousers, helmet, gloves and boots) is also essential. With safety in mind, it is well worth purchasing or constructing a device to hold the logs whilst cutting.

Log size is also about the width of the log so anything over 6in (15cm) will need splitting. A good, steady chopping block is essential, set at a comfortable working height. There are two main types of splitting axe: a splitting maul, which is heavy and blunt and relies on the weight and a good swing of the axe to force the log into two; the second axe is a more traditional shape with an edge that you can sharpen. It will cleave the log with less effort but can get it stuck in a reluctant knotty log. If it does get stuck, strike the handle of the axe with a sharp downward blow, avoid picking up the log on the end of the axe as that is a speedy route to a bad back.

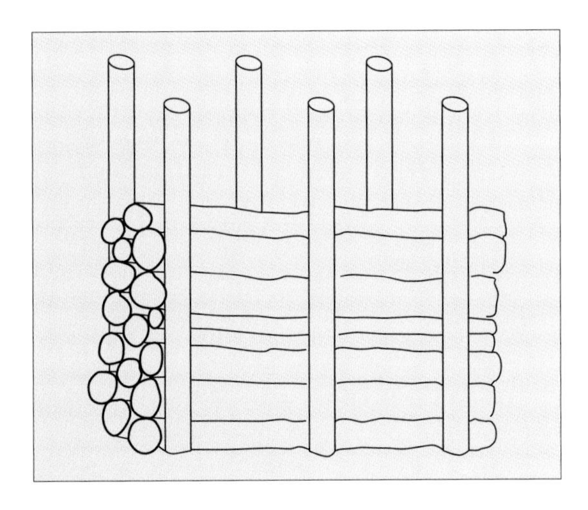

*Rack for crosscutting firewood with a chainsaw.*

A sturdy wide-legged stance is needed to aid accuracy, and should you miss the log, the axe will swing down between your legs and not bury itself in your foot! Wearing steel toe-capped boots is recommended to reduce the risk. Larger logs may need to be split using wedges and a wooden maul. A handless axe head is ideal to get started, as it will bite into the wood. Study your log carefully and choose to split it along an existing crack in the wood that is running through the core of the log.

Hand-cutting and chopping is good, honest labour and satisfying as the log pile grows, but if you are relying on wood as a fuel or logging on a bigger scale for sale, then you will be tempted to get mechanized.

*Chopping block with two types of splitting axe.*

**Options for mechanized firewood processing**

| Machine | Task | Pros | Cons |
|---|---|---|---|
| Chainsaw | Cross-cutting. | Fairly cheap to buy but much of the cost is in buying essential safety gear. | A lot of waste sawdust and noise and fumes. |
| Circular saw bench | Cross-cutting. | Much less waste wood. | Very noisy. Probably responsible for more lost fingers than anything else in forestry. |
| Log splitter (hydraulic) | Splitting lengths can split short or long, depending on the size of the splitter. | Much less effort than a hand axe!<br>Can be used to split long lengths prior to cutting. | Another machine to maintain. |
| Log splitter Screw type | Splitting chunky logs (short and broad). | Less initial outlay. | A bit slow and scary. |
| Firewood processor | Cutting, splitting and moving logs. | Does everything. | Expensive – you have to make it work hard to pay back the investment. |

A firewood processor is a combination of the cutting bench (either with a chainsaw or a circular saw blade) and a log-splitter; the log drops down and activates a hydraulic beam, which pushes the logs against a splitting blade. The splitting blade can be lifted to split four ways, dropped to split two ways or removed to allow round logs through. It also has a conveyor belt that takes the logs and drops them into a trailer or pickup. It can be run off the tractor PTO or have its own motor.

### *Final Seasoning*

The final stages of preparing your firewood should be to stack the chopped logs near the house in a shed with good airflow through it and a roof to keep the rain off, to become thoroughly air dry. Then, last but not least, bring it in near your stove to reduce the MC yet further. If you can live with a stack by your fire of a month's supply, which you then rotate, you will have the perfect firewood system.

*Firewood processor mounted on a tractor.*

*A woodshed near the back door.*

### *Kindling*

To start your fire you will need some form of kindling. This could be brushwood that you collect when you are felling your timber and bundle up as 'faggots', to dry out until they are so brittle you can break them into length by hand. Or you may prefer to have chopped kindling, which is a little more labour-intensive but makes quite a satisfying sight when stacked ready for use. If you are considering adding kindling to your firewood business, then think again. You will soon want to invest in a kindling machine, which will take dry logs in one end and produce perfect nets of sticks the other. Then you will spend the rest of your time keeping it busy to justify the expense.

**Kindling tip**

Alder wood is good for kindling, once chopped up it dries out very quickly. A green log can be ready for burning in a month or so, however almost any species soft or hardwood will do.

To supply a household's requirements for kindling, hand-chopping will be fine. You can speed up the task by cleaving the log with a froe, first one way to create slices and then the other to produce pegs just under 1in square (2cm) with a band around the log to prevent them from falling apart.

*Kindling being sold in returnable bags.*

## *Wood-Fired Boilers*

### Wood Chips

The idea of chipping wood for use as a fuel has been around for several decades already, but it is only in the twenty-first century that boilers that use this fuel have begun to be installed in Britain in any numbers. The best and most reliable boilers are all made in Europe, most commonly in Germany or Austria. Technology has come a long way in the last ten or twenty years, and these boilers are now incredibly clean and efficient. They are easy to light – sometimes with an electronic ignition, and produce very little ash.

However, although the idea of heating your home with chips that come from your coppice is a simple one, it can be very difficult to implement successfully. Chip-fed boilers depend on an auger to feed the combustion chamber and this means that if the auger gets stuck by a chip that is the wrong size, it stops working and the boiler goes out. Therefore, your chips need to be produced by a chipper that makes the right kind of chip – one that is within a fairly tight size specification. These chippers cost several tens of thousands of pounds and need a large tractor to power them. In addition, the chips need to be the right moisture content (the wood normally needs to be dried for 12–18 months before chipping). Then you need enough room for an under-cover storage hopper. The boilers work best with a very large accumulator tank to store your hot water. All this means that chip-fed boilers work better at a large scale, and are best at heating large public buildings that need a lot of constant heat, such as town halls, municipal swimming pools, hotels and conference centres.

*Wood chipper in action.*

*A log store for a wood-fired boiler at the Greenwood Centre.*

**Wood Pellets**

Pellets are usually made from dried, waste wood, which is a by-product from a sawmill or wood-using factory. So, you won't able to make them from the coppice you manage. There have been some concerns about their sustainability, as there is quite a lot of embedded energy in them and, until quite recently, they were mainly imported, adding to the environmental questions surrounding them. However, there are several advantages – they can be purchased in a bag and poured into the hopper, and there are no problems about their size or moisture content. Burning wood pellets is a lot better than using fossil fuels.

**Logs**

Perhaps a better solution for someone who has a ready supply of logs is a log-fired boiler. Of course, there are similar disadvantages here to the wood-chips option, in that you still require substantial storage space for the logs and a large hot-water accumulator tank; most houses will need a small building to put this in! However, as long as your logs are dry (ideally no more than 18 per cent moisture content and preferably a little lower), these are also very efficient pieces of technology, and your logs can be all sorts of shapes and sizes. Usually, log boilers are lit about once every other day, depending on circumstances, and the water in the big tank should provide all the hot water and heating you need for this time.

## CHARCOAL

Charcoal is also a wood fuel but the preparation is more elaborate and involves burning or heating the wood in clamps or kilns to remove the volatile components and to leave just carbon. One definition of charcoal has been stated as 'the residue of solid organic matter, of vegetable or animal origin, that results from the carboniza-

*Charcoal burning in an 8ft (2.5m) kiln.*
*(Photo: Mike Carswell)*

tion by heat in the absence of air at a temperature above 300°C' (Emrich 1985). I am not going to give you the recipe here for carbonizing animal remains but focus on the production of barbeque charcoal from hardwood trees. The advantages of charcoal over wood for cooking is that the carbon is a very stable product that burns easily and at a steady (hot) temperature without smoke and is ideal for quickly producing a bed of glowing coals for cooking over. Barbeque matters are regulated by British Standard BS EN 1860 – 2.

## History

Charcoal use goes back over 5,500 years (Kelly 1986), being used as the fuel for smelting in the Iron and Bronze Ages. It is possible to achieve temperatures of over 1,000°C and from that, metal can be produced from ore and worked into tools and, of course, weapons. In the last millennia there were four industries developed that depended on charcoal: iron, steel, glass and gunpowder. It is no accident that it is the areas where these industries dominated, like the Weald, the Forest of Dean and South Cumbria, that to this day have a high percentage of tree cover. Even though charcoal burning was blamed for deforestation, it would seem that coppice production and exploitation were largely kept in balance (Edlin 1973). By the beginning of the twentieth century most industry had found alternative fuel, such as coal and, later still, oil, but it is thanks to charcoal that humankind has progressed to where we are today.

## Species

Hardwood trees are the best for making charcoal. Some are less favoured – small-leaved lime in any quantity seems to be very reluctant to burn, and willow produces a soft charcoal, which makes it ideal for drawing charcoal but too crumbly for decent barbeque charcoal. Alder is a very hot charcoal and was used in the past for gunpowder production but it is a lightweight charcoal, which might be a consideration if you are selling charcoal by weight. All the other species have their own characteristics so, in an ideal world, you would have separate burns for each species. In reality a mix is more practical and seems to make very little difference to the efficiency of the burn. Some people have been worried about using rhododendron as a charcoal due to the possibility of its natural toxins finding their way into the charcoal. However, any toxins have been driven off or broken down during the charcoal-making process and, in fact, rhododendron makes a good charcoal.

## Getting Started in Barrels

The easiest way to make your first charcoal, without too much outlay, is to use a 45-gallon oil drum. These can be obtained from a variety of sources: farm feed suppliers may have empty molasses or machine oil barrels; car parts' companies or garages have them for engine oil; or try

*The four stages of making charcoal in a barrel.* Clockwise from top left: *A charcoal barrel that has just been emptied and ready to light again; charcoal making in a barrel; barrel burn showing the crossing wood that keeps the logs from falling out; the charcoal barrel sealed around and left to cool.*

your local scrap yard. Make sure that the barrel is complete without damage or rust holes and that the bungs in the end are of metal not plastic. Be sure that the barrel has not been used for storing any volatile compounds like petrol or paraffin. The first stage is to cut a slot in the side of the barrel and the easiest method is to use an angle grinder with a metal-cutting blade. You can hire these, though if you are just cutting one barrel then it might be better to seek out a garage or blacksmith who could do it for you. The slot should be 9in (18 cm) wide. I would recommend that you cut it to 7in (20cm) and fold under 1in (2.5cm) so that there are no sharp edges. Do this with a heavy pair of pliers and a lump hammer to flatten it. It is possible to cut the barrel with a lump hammer and bolster but it is very noisy and not so neat.

Once you have your barrel it is a good idea to light a fire in it (beware of vapours) and burn off any unwanted oil or molasses and the paint that will be on the outside. Not a very pleasant process, so don't do it in your back garden in vicinity of your neighbours' washing.

One of the ironies of making charcoal is that it is always good to have some charcoal to hand to start the burn. But this is not essential, if you can collect enough dead wood to act as kindling. Ideally though I put a shallow bed of charcoal in the bottom of the barrel and, using one firelighter broken in half, start two fires equally spaced in the bottom of the barrel. Purists amongst you will blanch at this heresy and please feel free to collect tinder and ignite it with some deft strokes of a flint. Being a pragmatist this author never has dry newspaper to hand in the woods so finds a box of firelighters and a lighter to be the charcoal burner's best friends! Once you have your fires alight in the bottom of the barrel, keep the barrel slot upright, unless you want to introduce a bit more oxygen, to get things going, by turning the slot into the wind. You may need to fan your fire, which will help if there is not much airflow in the barrel. The intention is to gradually build up the fires without swamping them at this stage, so keep adding dry wood (when you have had a burn you will have a supply of brown ends or part-burnt charcoal) and maximize the airflow into the barrel, as described. This first stage is known as the 'free burn' and is just about building up heat in the barrel. Spread the fires out when they are really going well and make sure they cover the entire bottom of the barrel. It will take at least an hour to get a good 2 or 3in (3–5cm) bed of glowing coals and plenty of flaming wood.

When you are satisfied that you have a really hot bed of fire in the barrel, then you can start filling the barrel with logs. Cut your logs to half-barrel length and make sure they are no more than 5in (12cm) wide – if they are bigger they will need splitting. Pack them into the barrel trying to eliminate air spaces as you go. This can be overdone, as some air is required to keep the fire burning. As long as your base fire is hot enough, you can fill the barrel right to the top at this stage and leave it to chug away for at least an hour. The main feature of this stage is clouds of billowing white smoke, which is in fact mainly water vapour as the wood is heated more and more, then the smoke turns brown and takes on an acrid taste. This is the volatile oils and tars in the wood being driven out by heat, a process know as distillation. There are more than forty different compounds in wood, which all can be removed by heating. Eventually though the carbon itself starts to burn and the charcoal burner's art is to stop that from happening. In the barrel the method is to reduce the oxygen by adding more wood. As the original wood in the barrel is heated and begins to burn, it shrinks; so by the time there are flames licking out, there will be space in the top of the barrel for more logs. Topping it up with wood takes the process back to the beginning again with clouds of steam and no flames. The magic thing about it though is that all the wood that is already charcoal in the barrel is just kept in a static state, as there is little air getting to it; so it continues to build up until the time comes to stop the process completely.

How long does it take to make charcoal? What a good question and how unanswerable it is. If you spend an hour getting your barrel hot and an hour cooking your first batch of logs you could then shut the barrel down and have a bit of charcoal in the bottom of the barrel. If, however, you do as suggested and top the barrel up and give it another hour, then you will get more charcoal. You can keep this up until the barrel is so full of charcoal that it is almost impossible to suppress the flames by adding more wood, which in our experience takes about eight hours. A colleague kept a 45-gallon drum going for 36h but it is possible that somewhere along the line they were losing charcoal to ash.

When it is time to shut the barrel down, you will have to roll the barrel over. This is where the design is so fabulous (credit for this must go to Richard Edwards who set up the Coppice Association in the 1990s; he explained to the author how to make charcoal this way). First dig yourself a shallow trench a little longer and a little wider than the slot in the top of the barrel. Then place some sticks across inside the barrel to stop the wood flopping out as it is rolled over. Position the barrel the exact distance from the trench and, with a deft boot or gloved hand, roll the barrel over so that the slot is exactly underneath in the trench. Smoke will be escaping all around but use the earth from the trench to press in under the barrel until no more smoke can be seen.

When you are satisfied that there is no smoke escaping, then you know that there is no oxygen getting into the barrel and without oxygen the fire will be extinguished. There is still a lot of heat and the final crucial stage of charcoal burning is allowing it to cool entirely. With a barrel it might take 3–4h. Large quantities will take longer and weather conditions also have a part to play. If, however, you get impatient and peep too soon, then the chances are that the heat and the air will re-ignite the charcoal and, before you know it, you will have a lovely pile of ash. Dowsing with water is the only remedy for this but it does rather spoil the charcoal if you get it wet.

Cool charcoal that has been cured (exposed to the air for 24h), can be graded and bagged, of which more in the next section.

### *Making Charcoal on a Larger Scale*

It is possible for one person to run eight barrels and, burning them over an 8h-day, you can get an average of 10kg (22lb) per barrel and so potentially 80kg (176lb) a day.

The major advantages of making charcoal like this are low overheads and flexibility, as you can take the barrels anywhere with relative ease. The major disadvantage is that you are working in amongst the smoke all day and it can be quite unpleasant. Eventually, if you have started to develop a market for your charcoal, then you will want to progress to a charcoal kiln.

### *Charcoal Burning in Kilns*

There are many varieties of kilns ranging from 4ft (120cm) wide to 8ft (240cm). They consist of a ring of steel with a rim welded on. The ring sits on box-shaped ports and has a lid that is closely fitting and so can be sealed to make it air-tight. Some have a second tier that can be stacked on top to maximize the amount of wood to be burned.

*An empty charcoal kiln. (Photo: Mike Carswell)*

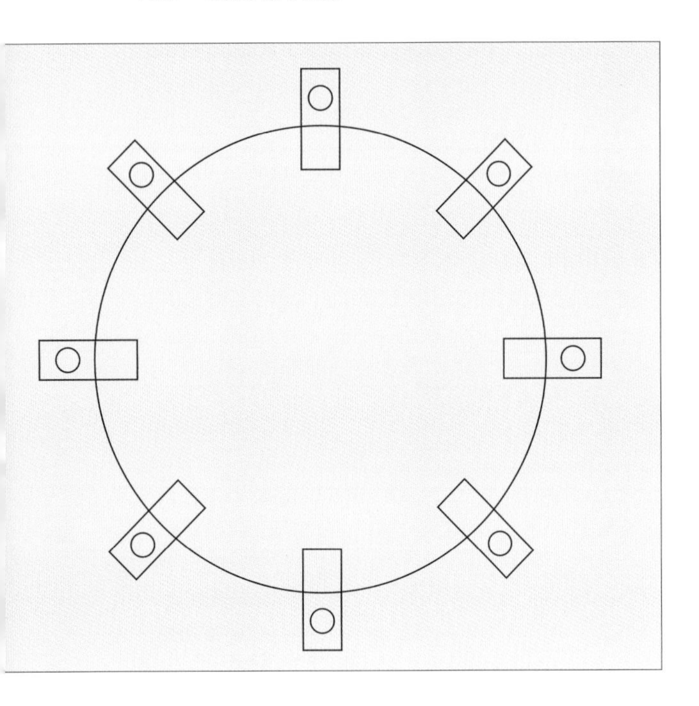

*Position of ports on a four chimney kiln.*

*The first logs laid out like the spokes of a wheel (Photo: Mike Carswell)*

## Preparation

You need a level site at least 6ft (180cm) wider than the kiln diameter. The ground needs to be sound and free of tunnels or potential air sources from underneath. It is advisable not to site your kiln on peat, as the fire can penetrate down into the ground. A supply of soil is important and, where the natural soils are sparse, you may have to import some from nearby. Make sure there are no trees near that will be scorched or smoke damaged and that there is an even airflow around the kiln; but be prepared to shelter the kiln from winds if it is at all exposed. First lay out the ports, spacing them evenly. Small kilns may just have six but larger ones have eight. Lay them like the points of the compass with the chimney holes on the outside.

With help, move a couple of the ports out of the way and lower the ring down until it is sitting on top of the ports. Re-position the ports so that they are evenly spaced and protrude into the kiln just 3 or 4in (7–10cm). Now fill the gap that is between the ports, at the base of the ring, with soil and compact it down to create an airtight seal. This leaves just the ports themselves as channels for the air to enter the kiln at the bottom. You are now ready to fill the kiln.

## Filling the Kiln

The first stage of filling is to create a platform of wood at the bottom of the kilns, so that the air is channelled from the ports to the centre. Start by laying out logs at least 4in (12cm) thick, no more than 6in (15cm), one each side of the ports and pointing to the centre, like the spokes of a wheel. Then place logs crossing these spokes so that they bridge the air channels and form a platform with a hole in the middle. Bridge the hole in the middle with some dead sticks, which will act as kindling and maintain the air space underneath, then pile on some charcoal or brown ends; if you really are at the very beginning, just make do with dead wood, but it must be dry. Stack this up in the middle of the kiln, as it is important that the fire starts in the centre and works its way up and out. Do not light it yet – first stack the kiln with wood. A simple statement but amazing how many variations there are on this particular theme. The author's preference is for a fast and furious version, where the wood is not placed but just tumbled in with a token bit of moving the logs around to minimize large gaps. There are plenty of burners around who will tell you that you must position the logs carefully to maximize the amount that you are getting in and minimize the air spaces. Our experience is that it is possible to exclude too much air, making it harder to get the kiln up to temperature. The key element to profitability is time, so don't get too obsessive about kiln stacking.

The same rules apply to the size of logs that you put in your kiln. They will need cutting to length and you can refer back to cutting firewood for suggestions ranging from bowsaws to firewood processors. The smallest diameter logs of 2in (5cm) can be up to 3ft (1m) long; logs up to 6in (15cm) should be cut to 2ft (60cm). If they are bigger in girth than 6in (15cm), please don't spend a lot of energy chopping your logs, just cut them shorter. You can charcoal logs up to 18in (45cm) diameter but you will have to slice them down to 9in (25cm)long.

Fill your kiln until the logs are right up to the top of the kiln, then add more until it is a good 12in (30cm) above the top of the rim; then balance the lid up on top of the stack. This will allow the hot air to flow out from under the lid, therefore drawing more air in through the ports.

*A kiln full of logs waiting to be lit (Photo: Mike Carswell).*

## Lighting the Kiln

Perch the lid on top of the stacked timber. Take a straight stick and pick a port, try poking the stick in through the port to check if the way to the centre of the kiln is clear. Then, removing the stick, tie a rag to the end and dowse it in paraffin (do not use petrol or anything more volatile). Put the paraffin can a safe distance away then light the rag. Push the burning rag on the end of the stick right through to the centre of the kiln.

You will soon see smoke emerging from under the lid, once it gathers momentum there is no danger of it going out. Occasionally, if your kindling is damp or it has been raining heavily when filling your kiln, then you may have to have a second go at lighting it (try a different angle through another port).

This is the moment that you can sit back and take a breather because the next 1–2h are the 'freeburn' and there is nothing to do but marvel at the quantity of smoke that one kiln can produce.

As the wood in the kiln gets going it will start to shrink and the lid, which at first is perched up above the rim, will start to sink down. Keep an eye on it at this stage because it is possible, if the burn is uneven, for the lid to slide off to one side, which can be somewhat scary and hot to rectify. It needs to come down gradually to sit within the rim. After an hour or so, there is so much heat that the wood gives off 'woodgas', which is volatile and ignites with a flare and a roar. Quite impressive when you light the kiln in the evening and the daylight is fading. It can suddenly shoot out of the ports and singe your ankles, so do wear stout boots and gauntlets to protect your hands and arms. Flaring is a good sign and indicates that the kiln is up to temperature. Don't over do it though, as it does seem to have a detrimental effect to burn off too much woodgas at this stage. If you have over-filled the kiln so that the lid takes too long to settle down on to the rim, this can be a problem. The kiln will get very hot at this stage but overall the burn is cooler and slower. Judging just the right amount of wood for your kiln and conditions is part of the endless fascination of charcoal burning.

*Kiln just lit.*

Once the lid is settled down all around, you can seal the gap with sand. At the same time, place your chimneys onto every other port and seal the gap at the bottom so no air can get in but the smoke will be drawn up the pipe. This encourages a circulation though the kiln, where the air is drawn in through the only remaining air inlets and feeds the fire in the centre of the kiln and pushes the heat up and round and down and out through the chimneys.

Hopefully, this creates an even burn right through the kiln. You can leave the kiln at this stage though it is a good idea to be somewhere close by. But it is perfectly alright to have a good night's sleep and then in the morning change the chimneys around one place, so that the inlets become chimneys and vice versa. The 8ft (2.5m) kilns photographed (see page 152) take almost 24h to burn, smaller kilns will take less time, so you will need to experiment and not leave them unattended until you are confident how long it will take.

*RIGHT: Lots of white smoke (steam) as the free burn gets going. (Photo: Mike Carswell)*

*BELOW: The chimneys in place. (Photo: Mike Carswell)*

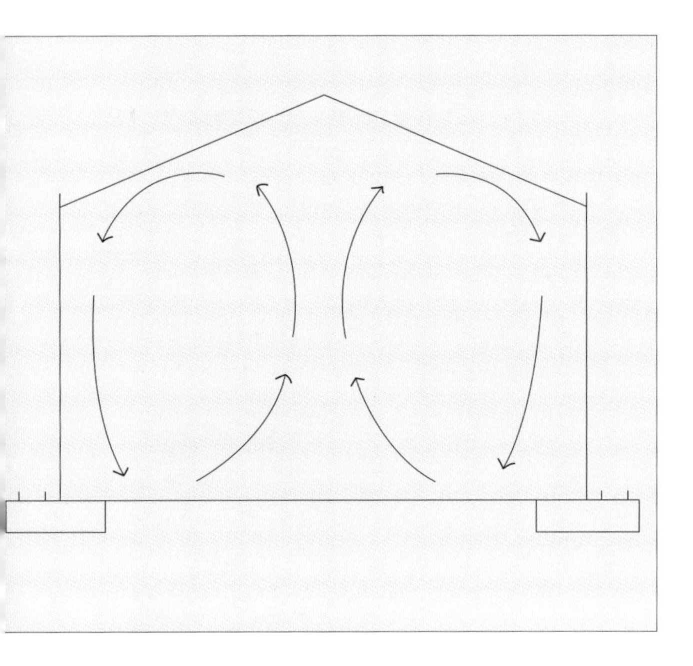

*Heat circulating in a kiln.*

*The kiln in the foreground is very nearly ready to shut down. (Photo. Mike Carswell)*

How do you know when the kiln is ready to shut down? Well it is all in the smoke. You know that in the first instance the smoke is white and mainly water vapour, and as the wood heats it becomes brown and acrid as the oils and tars are driven out. Then, as the carbon begins to burn, the smoke becomes blue and wispy, and at this stage you need to shut the kilns down. In reality it is a gradual process and in fact an hour or two more or less is not a problem. Too soon, the smoke is hard to suppress and there will be a lot of 'brown ends' or unburned wood; too late, there will be a higher proportion of ash and less charcoal.

## Shutting the Kilns Down

Shut the kiln by taking off the chimneys and filling all the port holes with socks filled with sand (rabbits) and making sure there is no smoke emerging from any gap anywhere; use soil or sand to block it all off. Check around the rim for gaps and around the base. When you are confident that no smoke is coming out, so no air is getting in, then you can leave it to cool. A ring kiln will take at least 20h to lose the heat completely. Do not open it too soon, as the residual heat in combination with the oxygen will re-ignite the kiln. The tell-tale sounds are a faint crackling sound, a little bit like 'rice krispies' when the milk is poured over. If this happens, dowse with water as quickly as you can, as the fire will soon spread through the kiln. Best practice at this point is to cure the charcoal by exposing it to air for 24h before bagging. By burning the kilns just once a week you can be confident that the charcoal is truly cold when you open the kiln and already cured. Problems arise if you try to burn the same kiln twice in a week, as the turnaround is quite tight and you will have to empty the kilns into sacks to cure, and then into bags for sale.

## Grading the Charcoal

The cold charcoal will need grading on a riddle to make it fit for use on a barbeque. The standard riddle is a ¼in (10mm) mesh to remove the ash and dust and a ½in (15mm) mesh to remove the charcoal fines. These should be saved for using as a soil improver or compost constituent, or used for paths or burned in closed stoves. The remaining charcoal is picked over to remove 'brown ends' and then either stored in sacks or packaged into bags for sale.

*Using a rotary charcoal grader.* (Photo: Mike Carswell)

**Charcoal packaging options**

| Bag | Advantage | Disadvantage |
|---|---|---|
| Brown paper | Cheaper<br>Aesthetics (if you prefer the style)<br>Available in small quantities | Dirties easily<br>Gets damp |
| Glazed paper | Wipe clean<br>Colour printing | More expensive<br>More manufacturing<br>Also spoils in the damp |
| Polythene | Can be stored outside<br>Choice of styles | The charcoal can sweat in hot weather |

### Packaging the Charcoal

The best way to sell charcoal is by volume and a bucket measure is a good way to package a standard amount. Whatever packaging you are using, it is most important that you keep the bags clean and staple or stitch them neatly. A clean, smart bag is a great selling point.

Bags are usually made of paper but can be plastic – both have advantages and disadvantages.

The best selling point that you can have is that the charcoal is locally made, so it is important to emphasize that on the packaging. Take the opportunity to spell out the ecological benefits of supporting woodland management. Most important though is to emphasize the quality and ease of lighting. We cannot compete with imported charcoal on price, so it is essential that people appreciate the quality of British charcoal, and if you can get them to try it once, they will come back for more. Getting your own bags printed is quite a commitment with minimum print runs sometimes into the thousands, so accessing bags through a cooperative scheme, such as is available through the Coppice Association North-West, is a good option; alternatively approach an existing burner and buy some bags from them.

Store your charcoal in dry airy sheds, as it is important that it does not get damp. It is, by nature, hygroscopic, which means that it will take up moisture from the air like a sponge, compromising the quality of the charcoal.

**Buying British**

Buying British made charcoal is best because it:

- Generally is of better quality, has fewer residues and is easier to light.
- Cuts down on 'air miles' and unnecessary transportation.
- Is easier to be certain that the timber has come from a sustainably managed source.
- Supports local woodland businesses, benefiting both people and wildlife.

### *Earth Burn*

The traditional way to make charcoal was in a clamp. The method described was the one used in the north of England. Timber was stacked in a very orderly manner around a central post or 'motty peg'. The stack was dome-shaped and when complete was covered in grasses, bracken and turves, and then a layer of sieved soil. The clamp was lit by pulling out the central post and pouring burning coals down into the centre. The burn was controlled by careful monitoring of the progress and suppressing any outbreak of

*The earth burn being opened by three generations of charcoal burners.*

*Charcoal fines being used on a path.*

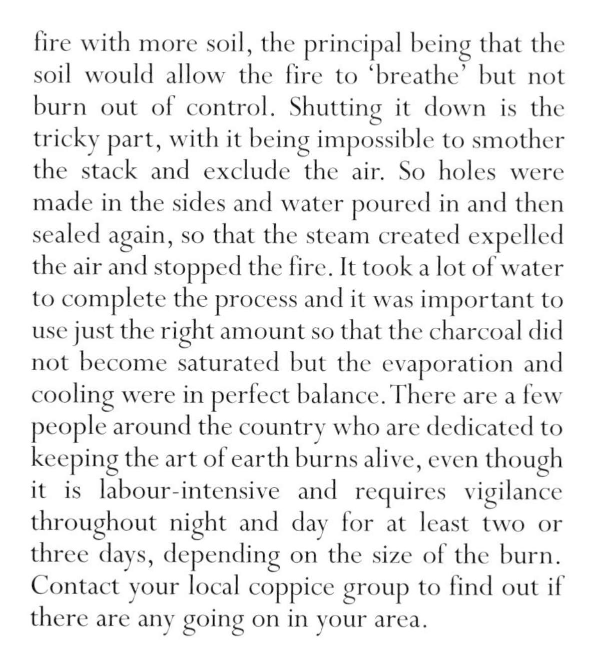

fire with more soil, the principal being that the soil would allow the fire to 'breathe' but not burn out of control. Shutting it down is the tricky part, with it being impossible to smother the stack and exclude the air. So holes were made in the sides and water poured in and then sealed again, so that the steam created expelled the air and stopped the fire. It took a lot of water to complete the process and it was important to use just the right amount so that the charcoal did not become saturated but the evaporation and cooling were in perfect balance. There are a few people around the country who are dedicated to keeping the art of earth burns alive, even though it is labour-intensive and requires vigilance throughout night and day for at least two or three days, depending on the size of the burn. Contact your local coppice group to find out if there are any going on in your area.

### *Charcoal Fines*

Fines are the small particles of charcoal that fall through your grader. Preferably discard the fines that come through the ¼in (10mm) mesh, as this is mainly ash, but gather up the fines that come through the ½in (15mm) mesh, as this has many uses, such as:

- Soil improver.
- Paths.
- Briquettes.
- Filtration.
- Slug deterrent.

#### Soil Improvement

Of the many uses of charcoal, the most important is its horticultural use. Charcoal is hygroscopic, in other words it attracts water from the air and soil and holds it within its sponge like structure. It also holds water-borne nutrients and keeps them available for plants to exploit. In addition, it opens up the structure of heavy soils and allows better root development. Best of all, it does not break down in the soil, so the more you put on the better your soil gets and will remain (sequestering the carbon) for hundreds of years!

# Chapter 9
# Setting Up a Coppice Business

## INTRODUCTION

So, you have begged, borrowed, bought or rented a wood and you are fired up with the knowledge of all the wonderful things that you can make and harvest from this precious resource. How do you go about making a living from it? Maybe your aim is just to make it pay its own way. Either way it is no longer an option to sit on this resource and do nothing with it, even if it is only to provide sufficient wood for your own use.

## CHARCOAL AND FIREWOOD – THE BACKBONE OF A COPPICE BUSINESS

What you can produce from your woodland will depend on the nature of your wood and the products that are available; however, you can guarantee that if the woodland has been neglected prior to your taking up the stewardship of it, then there will be a good deal of low-value wood waiting to be utilized, which is fit mainly for firewood and charcoal. These two products are often the core of a coppice business, as they provide year-round cash-flow. There is sometimes a lull between a wet August, when charcoal sales have plummeted, and October, when logs sales get into swing, and this is the time to plan an extended break to recuperate from the rigours of the work.

### *Building a Customer Base*

The main consideration before launching yourself into producing wood fuel is: can you keep up with demand? Are you going to build a customer base just to let them down in the future? Barbeque charcoal is a classic example of this; you have just bagged your first kiln of charcoal into shiny new bags and have loaded them into the back of your van but then have mixed results taking them round to garages and garden centres. Some of the traders have already got a local supplier; others are not convinced that they can sell your charcoal at a premium when they are selling imported charcoal at half the price (by weight not bulk). Eventually

*Having a barbeque under an umbrella.*

*Large-scale firewood delivery.*

someone decides to take a chance and buys ten bags. Buoyed up by your success you try a few more places and persist until you return home with an empty van, a full wallet and a smile on your face. So you now have a customer base of maybe six shops/garages – fantastic, but a word of warning because barbeque charcoal sales are entirely weather-related. Sometimes we have a mini heat wave around the spring bank holiday and suddenly all your customers are on the phone wanting twenty bags straight away. Have you managed build up a stock pile? Or are you waiting for the kiln to cool so that you can bag it up and get it delivered? The reality is that there are many weeks in the British summer when sales are very slow and, as charcoal is bulky to store, it is not always practical to keep making it in the hope that the good weather will return. However, luckily for charcoal burners there are always people keen to get outside and have a barbeque as soon as the weather turns (or under an umbrella, for the really keen), so be prepared.

Creating a customer base for firewood is a similar story to that of charcoal, except that now it is winter and you are waiting for the elusive cold snap that sends everyone scurrying to light the living room fire to supplement the central heating. As soon as that happens, sales will escalate and, unless you are going to disappoint people and lose good will, you will have to be poised to meet the demand.

The moral to this tale is to build your business slowly and plan ahead for all eventualities. Building a reputation for a courteous, efficient service will ensure you have lots of repeat orders and that your details are passed on by word of mouth.

### *Promotion*

Word of mouth may be the best way to grow your business but to get started, a card in a local shop will produce local customers and keep your delivery miles down. Do not get impatient and put a jazzy advert in a phone directory unless you are fully prepared for the sudden demand that exceeds your supply. Similarly a website may seem like an essential part of a modern business but you should decide whether you want to be fielding enquiries from America or perhaps posting charcoal from Cornwall to Inverness. Be clear about your geographical limits and reduce the impact of transportation.

District agricultural shows or countryside festivals are an ideal forum for meeting potential customers and getting yourself known. Often event organizers will want to charge you

*BHMAT apprentices demonstrating at the Westmorland Show.*

*It is more efficient to make components in batches.*

for a stall, but if you have a skill to show people, then you may be able to attend for free or ideally be paid to demonstrate. Sometimes it may be economically viable to pay for a stand but bear in mind – even if you have a lot of lovely items for sale and sales are good, takings are not the same as profit. Your time attending the show, as well as the time taken to make the products and the cost of the stand, must be covered in the prices that you charge for your items.

## Pricing

How do you know what to charge? If the product is something widely available, like firewood logs, then the easy option is to do a bit of market research. Phone around for quotes from local suppliers and get a feel for what others are charging. You do not want to whip up instant bad feeling by undercutting your competitors, but pitching your prices high may result in slower growth of your business. You need to be realistic about the cost of materials, overheads and your time. Be sure that you are comparing your products with an equivalent from your competitor, as some firewood merchants sell unseasoned wood and may be less scrupulous about their selection of species.

Charcoal pricing is more difficult: you cannot possibly compete on price with imported charcoal, especially when selling by weight. You can point out though, the obvious benefits of a 3kg (6½lb) bag of lovely, fully converted, home-grown charcoal, as opposed to the imported 3kg bag (which is half the size and contains denser charcoal, meaning it is harder to light). Ultimately, you are relying on the retailer taking a chance on the marketing advantages of a high-quality, local product. Once a keen barbeque fan has tried British charcoal, they will come back for more. So expect sales to grow when you have established an outlet. Again with charcoal pricing you will be best advised to go along with existing charcoal producers, if there are any in your area, or use the internet to gauge prices.

For more bespoke items, it may be best to work out the cost of materials, plus business costs (rent of workshop, running a vehicle, etc.), plus time. The only disadvantage with this is if you are scrupulously honest about your costs you can end up with products so expensive they will not sell. The only solution to this is to reduce the component costs. Can you cut the cost of the materials? Can you reduce your business costs? Can you make the products any faster? Reducing the time it takes to make something has the most impact on cost. When you are starting a business, products take longer to make, but with practice you can find ways of speeding up. If you are making a run of identical items, then do each process in a batch rather than completing a product before starting the next.

As craftspeople we are often being exhorted not to undersell ourselves. It is important to value your time and skills and expect a decent standard of living. However, when you are establishing a business that is largely craft-based and manual, you do have to be realistic. Many people find it easier to establish themselves prior to taking on the financial pressures of children and mortgages. They keep their overheads low with

*A discrete bender blending into the woods.* (Photo: Sam Ansell)

*Woodland Pioneers with the products that they have made over the week.* (Photo: Paul Spencer)

informal housing (caravans, yurts and benders) and, most importantly, keep borrowing to a minimum. Taking out a loan for a shiny new vehicle may be attractive but the benefits of having reliable transport will have to be balanced against the pressure of repayments. Investing in machinery that will make the job more efficient should be a priority, but look into the availability of business grants and loans that will be less of a drain on the business resources.

## TRAINING

Some people are great at teaching themselves; they get a good book on the subject, read up and then get cracking. However, this can lead to a lot of 'learning by experience', 'trial and error' and 're-inventing the wheel'. Some judicious training can be a shortcut through this sometimes painful process. Courses provide an opportunity to really pick the brains of someone who has had years of experience in your chosen new field.

Courses vary from taster sessions of a few hours to a few days or more. They have the advantage that all your tools and materials are provided – you just need to turn up with a thirst for knowledge and new skills. A note of caution though – attending a week's course may boost your confidence and fire your enthusiasm but it does not mean that you can immediately start producing craft items to a quality that is saleable. There is no substitute for practice and, if you are serious, you may well need to repeat courses or seek out advanced tuition to bring your own skills level up to scratch. You will definitely need to set aside time to make your chosen products for your nearest and dearest first, as they are, generally, less insistent on perfection!

There are a few opportunities for extended training. The Green Wood Centre at Ironbridge, Shropshire, run a 5-day, Open College Network (OCN) certified course in coppicing, as well as a number of other related courses. Individuals such as Ben Law in Sussex and Mike Abbott in Herefordshire offer 6-month placements for one or two individuals. The Bill Hogarth MBE Memorial Apprenticeship Trust (BHMAT) offers a 3-year apprenticeship in coppicing, with an accredited diploma. Placements are based mainly in the north of England. To be eligible for the apprenticeship people attend the 'Woodland Pioneers' Introduction to Coppicing week held each September in South Cumbria. The Bill Hogarth Coppice Diploma is also available as the National Coppice Apprenticeship (NCA) lead by the Green Wood Centre (Small Woods Association). The NCA is piloting a new Forestry Apprenticeship, which will have a 'Coppice Pathway' and will be a 2-year training for 18–25 year olds.

## BUSINESS SKILLS

### *Book-Keeping*

As soon as you decide to make a business of your woodland activities, then it will be necessary to submit self-employed accounts to the Inland

## Sample spreadsheets for expenditure and income

### Expenditure

| Date | Item | Cash | Bank | Mats | Motor | Admin | Advert | Legal | Pension | Drawings | Capital |
|---|---|---|---|---|---|---|---|---|---|---|---|
| 5 April | Fuel | 50 | | | 50 | | | | | | |
| 7 | Ins | | 146 | | | | | 146 | | | |
| 8 | Stamps | 6.24 | | | | 6.24 | | | | | |
| 8 | PO | 5 | | | | | 5 | | | | |
| 9 | Timber | | 300 | 300 | | | | | | | |
| 14 | Glove | 4.70 | | 4.70 | | | | | | | |
| 14 | Fuel | | 50 | | 50 | | | | | | |
| 19 | Van | | 2000 | | | | | | | | 2,000 |
| 21 | Tools | | 86 | 86 | | | | | | | |
| 21 | Fuel | | 50 | | 50 | | | | | | |
| 22 | Car tax | | 160 | | 160 | | | | | | |
| 22 | Tel. | | 20 | | | 20 | | | | | |
| Total | | 60.94 | 2812 | 390.7 | 310 | 26.24 | 5 | 146 | | | 2,000 |

### Income

| Date | Item | Cash | Bank | Char. | Teachg. | Demos | Coppice | Contracts | Firewood | Total |
|---|---|---|---|---|---|---|---|---|---|---|
| 5 April | Jones | 60 | | | | | | | 60 | 60 |
| 8 | Smith | | 240 | | | | | 240 | | 240 |
| 9 | Garage | 88 | | 88 | | | | | | 88 |
| 13 | SLDC | | 350 | | 350 | | | | | 350 |
| 17 | Taylor | 60 | | | | | | | 60 | 60 |
| 17 | Benson | | 60 | | | | | | 60 | 60 |
| 21 | Hurdles | | 145 | | | | 145 | | | 145 |
| 25 | Garage | 88 | | 88 | | | | | | 88 |
| 29 | Sticks | | 50 | | | | 50 | | | 50 |
| 29 | WI | 40 | | | | 40 | | | | 40 |
| Total | | 336 | 845 | 176 | 350 | 40 | 195 | 240 | 180 | 2,362 |

Revenue (IR). This may seem a bit daunting, but need not be too scary. Basically, it is a matter of keeping a record of what you earn and what you spend in relation to the business. There is plenty of help and advice out there. A good place to start is Business Link (see Useful Addresses), which can give free advice to new businesses. The Inland Revenue (IR) is also a useful place to get guidance; after all, it is to the IR that you will have to submit your business accounts, so you may as well keep your books in a manner that will suit them. You do not necessarily need an accountant as you will have to keep your books yourself anyway and submitting them to the IR is fairly straightforward. But bear in mind that an experienced accountant with up-to-date knowledge of regulations about tax allowances can prove cost-effective, once your business is established.

When you are starting out all you require is a basic accounts book, available from a stationers, with columns for 'ins' (income) and 'outs' (expenditure), and a running total for the end of each week or month. Jot down your expenses and file the receipts, as you need to keep the proof of purchase. Keep your expenses in separate columns, such as 'motor costs' and 'materials', and that way you can claim all your materials costs against tax but maybe only 50 per cent of your motor costs, if you also use your vehicle for private trips. A note book kept in the car and filled in at the start of each journey will help you work out the percentage of business use. Likewise with income you can keep it all in one column or separate it out into 'firewood' and 'charcoal', etc. in order to keep tabs on how different aspects of your business are doing.

If you are computer literate, compile your accounts on a basic spreadsheet. This will enable you to quickly add up your columns and potentially do more advanced calculations, depending on your level of skill.

*Coppice worker's trusty aid.*

### *Business Plan*

At some point, sooner or later, especially if you apply for a grant, you may be required to produce a business plan. This too can be kept quite simple and does not need to be something that you lose sleep over. At its simplest it is a plan of how your business will perform in the forthcoming few years. So you need to estimate how much income you expect to receive. If you are right at the start of setting up a business, this can only be an estimate but should be as accurate as possible. Put in actual contracts that you already have and realistic estimates of sales. Then take a long, hard look at your expenditures and fill those in as honestly as possible. A personal survival budget can be useful and should include all those hidden expenses and little luxuries that you are bound to indulge in sometimes. This is essential, as you must calculate your likely drawings out of the business. A sample form is in Appendix I. The Profit Plan (Appendix I) will help you to fill in all your business expenses and plot them against your income so you can get an idea of whether you are likely to be in profit or loss at the end of the year.

**The Business Plan**

A business plan may contain:

- Business proposal.
- Curriculum Vitae.
- Detail of the business.
- Analysis of the market.
- Price policy.
- Capital requirement.
- Profit plan.

A business plan should be a useful document that you actually use and refer to from time to time. It can be a good way of assessing the business against targets that you set yourself and, if necessary, adjusting the way the business runs. The plan should detail the nearest competitors and include the thinking behind your price-setting strategy. There's not much point in writing something that sits on a shelf gathering dust.

### *Ten Top Tips for a Successful Coppice Business*

1. Start your day early! The hour between 8 and 9am is very valuable in terms of getting a job done. You can always knock off early if you have everything under control.
2. Avoid getting into debt. Keep within a realistic business plan and be sure that you have the required cashflow before splashing out. Repayments can be too much of a burden on a new business.
3. Look after your body. Coppicing is a physical job and you are dependent on your strength and health to stay in business. So do not skimp on things that improve your performance, such as a regular massage, a yoga class or sufficient holidays to recuperate!
4. Be methodical. Keep records, set aside time to keep up with invoicing and book-keeping, or if you are hopeless at these things, consider getting help to set up a fail-safe system.
5. Network widely. Coppicing is a small industry but in general people are happy to help each other. So be open to skill sharing, cooperative working and generally being nice to each other!
6. Look after your tools. This particular author does aspire to be better at this! Keeping tools sharp and dry should be your basic aim. Old tools that have been lovingly looked after are infinitely superior to most tools you can buy new, so treat them with respect.
7. Health and Safety (H&S). This may seem like a boring irrelevance to some or an obsessive's preserve, but actually H&S is just common sense. Accidents do happen. Proper protective clothes and equipment reduces the seriousness of accidents. Avoid working when you are too tired or stressed, so that the likelihood of accidents is reduced. Being mindful of your colleagues and people around you will prevent accidents. Do a first-aid course and keep a properly stocked first-aid kit on site. Be sure to have valid insurance.
8. Look after the planet. We only have one, so it seems logical that we should focus on how we impact on it. How can you reduce your carbon footprint? Are you enhancing the environment or degrading it? What more can you do to address these issues in the way you run your business?
9. Do not spread your self too thinly. Tempting as it might be to diversify in order to capture new markets. Or become a 'new skills junky', unable to resist learning new crafts. There will come a point when you must consolidate your business and focus on what you are really good at or, perhaps, boringly, what is most profitable. Strike a balance between variety and what are proven to be your bread-and-butter products.
10. Enjoy yourself, whether it is at work or at play.

## ADDITIONAL INCOME STREAMS

### *Offering Your Site for Training*

At some point when you have established a woodland-based business, you will probably consider offering training. If you are uncertain that you have the skills yourself, then you can always bring relevant experts in to provide the training. At what stage should you consider offering this yourself? There are a lot of stories of people who have done a weekend course themselves one month and begin training others in those skills the next. This is considered very bad practice; from a participant's viewpoint you want to be sure that your tutor really does know what they are talking about. However, someone who has spent their life working in a trade may have immense skill but no idea how to teach those skills to others. Someone else may be relatively new to a subject but a natural communicator or trained in teaching and very able to draw out the best in their students. As long as you are honest about your abilities and, for example, do not try to teach advanced swill basket making when you have only made half a dozen yourself, then there is no reason why you should not pass your knowledge on to others. If you are new to training, then start modestly with a few friends gathered for a workshop on a subject that you are really confident about, and get feedback from your students on how to improve.

In order for your trainees to have the most positive experience, you should attend to the following points:

- Clear instructions to participants regarding directions, what to bring, start time.
- Good preparation with an organized and logical structure to the course.
- Plenty of good-quality materials.
- Sharp tools and enough to go around, so that no one is waiting too long.
- Sufficient cover in the event of wet weather.
- Keep the trainees busy, they have paid for a 'hands-on' experience. Avoid too much talking and not enough doing!
- Handouts with useful additional information, where appropriate.

The last point is very important, as most workshops that you offer in the woods are, by

*Students happy at their work.*

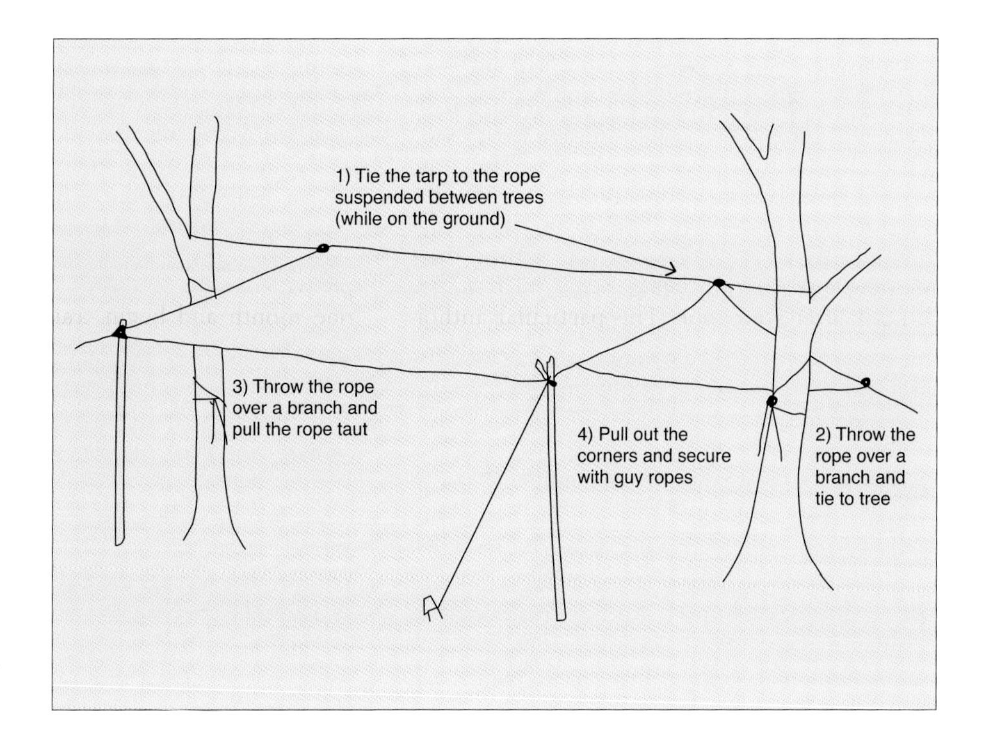


*Tarpaulin shelter for working under in the wood.*

definition, about practical subjects. The theory and history related to your subject may be fascinating but this information should be fed in amongst practical elements. Technical information, which may take a bit more assimilation, can be given out in printed form so that the students can read it in their own time.

### Suitable Sites for Training

If you are going to offer training there are a few things you need to consider:

- Venue.
- Parking.
- Shelter.
- Facilities.
- First-aid.
- Risk assessment.
- Insurance.

You do not necessarily have to own a wood or a workshop to host courses but, if you intend to hire or borrow a venue, you will have to ensure that you do have the owner's permission and that the site is suitable. The most obvious requirement is for people to be able to find the site and access it through a range of transport options. If they are driving: where will they park? How many cars do you anticipate arriving? Will they be able to get in and out in all weathers? Are there options for arriving on public transport and what arrangements should you make for collection from bus stops or train stations? The most basic facility is shelter. This can be as simple as a tarpaulin suspended over a rope that is strung between two trees and securely tied down. Good for most weather conditions, though it can be rather flappy and hazardous in a strong wind and constant rain will put a dampener on most situations unless your tarps are big enough to house everyone underneath comfortably.

Some form of toilet is essential, even if it is a temporary portable type or a kind of earth closet. Along with that must be hand-washing facilities and, needless to say, privacy.

Other extras that may be essential are tea and coffee facilities, and even hot food if the course is outside in the winter. All these things add to the feeling that your course participants are being cherished and cared for!

### First-Aid

It is a sensible precaution to have someone on site with a valid 'first-aid at work' certificate during any type of public event (especially if that event involves sharp-edged tools). A well-stocked first-aid kit should be kept somewhere safe but accessible, and with it should be an accident book to note any incidents, and details available for accessing the nearest Accident and Emergency department.

### Risk Assessments

Prevention is the best approach to accidents. Identifying risks and taking avoiding action should become second nature when dealing with the public. List possible hazards on a risk-assessment form (Appendix II), making a note of how likely they are to occur and how serious the consequences are of a potential accident. Then list the steps you will take to reduce the risk, such as providing protective equipment, cordoning off areas of danger and simply making people aware of the risks. They may be unfamiliar with the environment and unaware of hazards that seem obvious to you.

### Insurance

If all your preparation and foresight fails, then there should be the back up of knowing you have public liability insurance. This is in addition to the common-sense steps you need to take to avoid danger. Most policies will insist that you have a risk assessment in place for any event, and even a health and safety policy. Insurance need not be exorbitant – there are a number of membership organizations that offer insurance to members at very reasonable rates. These include the British Trust for Conservation Volunteers (BTCV) and the Association of Pole-lathe Turners (APT).

### Publicizing Your Course

When you have all the practical elements in place, what you need now are the participants. They may be contacted through word of mouth, though you will need a flyer or leaflet to give to people that has all the relevant information on it. Leaflets can be great to hand out at shows or at events. Posters too should be put up where your likely customers gather. Now, as you develop this angle of your business, a website would be really helpful. It does not have to be very sophisticated with online booking but should be well-designed and attractive to draw people in. If you do not have your own website try and get your course listed on other relevant sites. Plan ahead and give people plenty of notice and keep them informed with a letter of confirmation of booking and clear instructions for finding the venue and what they will need to bring. As with any aspect of your business, attention to detail will be repaid with satisfied customers who return again and again.

## *Contracting*

When you have established a coppice cycle in your own woods, you may consider hiring yourself out to work in other people's woods. There is an opportunity for skilled coppice workers to be paid to undertake contract work. Many conservation organizations are prepared to hire experienced teams who have enough knowledge of ecology to be trusted with management work on sensitive sites.

### Pricing the Job

Some experience is necessary to be able to price a job; for that you need to be able to judge how long the job will take, bearing in mind any possible problems. For example, bad weather can make vehicular access impossible, making it difficult to remove any material. The working day will be reduced if you have to carry your equipment any distance. If you visit the site on a sunny day in September, imagine how it will be on a snowy day in January with short days and problems lighting a fire, if the contract involves burning up brash. Think about how many people it would be practical to have on the job at one time for maximum efficiency. Take into consideration the size of the site and how many chainsaw operators could work, keeping a safe tree-and-a-half length apart. The minimum would be one person but you will have to calculate for two, as it is unsafe for chainsaw operators to work alone and your contract will certainly not allow it. If the contract involves burning the brash, then allow plenty of time for this as it takes roughly twice as long to clear up as it does to fell; this is the case even if you are just stacking firewood and not dressing out products.

When you have calculated how many people-days you think it will take, then multiply by a realistic day rate. Do not forget your hidden overheads, such as fuel, depreciation and office costs. This is where it can be difficult if you are competing against other contractors for the job. You are unlikely to know what they are charging and the temptation is to keep the price low to get the contract but you do not want to lose out, especially if you are paying people to help you. Just be as realistic as you can be and hope for the best.

*Contracting work need not be all slash and burn.*

*Coppice transport of the future?*

### Insurance, Risk Assessments and Health and Safety

You will certainly be asked to provide proof of your public liability insurance and sometimes it can be necessary to have up to five million pounds worth of cover. However, once you have a need for this level of insurance it is important that you get the work to justify having it. Employers' liability insurance is also a legal requirement if you are instructing others, whether they are paid or not. For either of these insurance covers to be valid all your chainsaw operators must have a valid chainsaw certificate, copies of which will be required to be shown to the insurer.

Risk assessments are often requested and a simple example of this is provided in Appendix II.

Your insurance company will insist that you have a health and safety policy and that you adhere to it. Again, this is largely a matter of common sense and it will help you to focus on what you need to do to keep yourself and your workers safe. An example can be found in Appendix III.

### Completing the Job to a Satisfactory Standard

Most contracts will involve a written description of the work to be done. Read this carefully and check any details that are not clear. A site visit with the manager is usual and is an opportunity to clarify the details. It is a good idea to take notes that will help you remember any extra verbal instructions, or take a recording device along if you find that easier. The importance of completing the job to the specification may seem obvious but sometimes it is harder than you would think. Keep in touch with the manager and inform them of any problems that arise. When the job is completed to everyone's satisfaction, submit a professional invoice promptly and await payment.

## CONCLUSION

For decades now the conventional forestry mantra has been that there is no value in our native woodlands. Timber prices remain low and the received wisdom is that there is no profit to be made out of our own woods. In this book we hope we have gone some way to exploding that myth. We may have to indulge in some creative accounting where in the 'income' column we include values for:

- Personal satisfaction.
- Creative fulfillment.
- True environmental cost benefits of local as opposed to global.
- Species'-rich environments.
- Human skill and knowledge.
- Preservation of heritage and continuity.

In the 'costs' column we must put:

- Lower standard of living than the average.
- Physical wear and tear of the human body.
- A bit of discomfort when challenged by the elements.

But just this simple accounting exercise shows us that the benefits of embracing a coppice based lifestyle far outweigh the disadvantages, as long as an easy life and easy money are not your top priority.

Lack of capital should not be a barrier to you progressing into a coppice career. As we saw in Chapter 2, 'Renting or leasing a woodland', (see pages 16–29) there are many ways to get yourself into the woods. If all you are ready to do is to tentatively dip in your toe, then get involved with your local conservation group and do some volunteering because that way you will soon find out if it really is for you or not.

Picking the right wood is crucial and 'reading the wood' will have furnished you with all the knowledge you need to decide where to put your energies.

Getting equipped for woodland work can be as basic as purchasing a billhook, a bowsaw and a bike (with trailer).

Coppicing can involve a lot of expensive machinery. In Chapter 3 (see pages 30–49) we explain how you can buy wisely, work out what you really need, and build up your tool store and equipment gradually as it becomes essential to your well-being or the expansion of your work. If you were tempted to skip the section on your legal responsibilities, go back and read it now – it will save you endless trouble in the long run.

Coppicing is something that can mean a lot of different things to different people. From a convenient and cosy way of selling 'clear fell', with no intention to foster the regrowth, to industrial-scale cropping of biomass, which has very little to do with biodiversity and environmental care. Chapter 4 (see pages 50–71) has described how coppicing is in fact a very specialized skill involving care for the rootstock, nurturing of the regrowth and informed decisions about timing of the cycles, with due regard to the habitat that you are managing and the overall health of the wood. Why? Because if you have become a committed manager of a coppice regime, creating habitat for all the fantastic variety of organisms that can live in your woodland is now your main aim in life.

Our most important message is that managing woodlands for wildlife and producing a coppice crop are both aims that you can achieve without compromising either. Your new found knowledge will govern your decisions regarding whether it is appropriate to coppice and, most importantly, how to coppice to meet all your objectives

Chapter 5 (see pages 72–87) illustrates how coppicing will reward you with a natural abundance that will amaze and delight you from the constant buzz of insects on a sunny day to the flash of wing of a woodland bird.

Chapter 6 (see pages 88–117) is dedicated to a tribute to the versatility and beauty of hazel. Wherever hazel grows, we have found a hundred ways of making use of it, and it is woven into the very fabric of our cultural history. Take it to your heart and let your imaginations be your guide.

How unfair then to condense all our other beautiful woodland trees into one chapter alone. Hopefully you will take the suggestions from Chapter 7 (see pages 118–141) and use them as a starting point for many more woody projects. If the idea of so much creativity is a little daunting, get stuck in with some more grounded jobs like feeding your fires and perhaps making some charcoal, as described in Chapter 8 (see pages 142–163). Nothing is more rewarding than fuel production because without that where would we be?

In our whistle-stop tour of our book we find ourselves back here at Chapter 9 Setting Up a Coppice Business. We do hope that you are inspired to attempt this on whatever scale – large or small. The rewards will be around you at all times and the earth will reap the ultimate benefit. It is only by putting all this knowledge into practice that we will achieve our ultimate goal of productive, sustainably managed woodlands that are good for people, good for wildlife and good for the planet.

# Appendix I
# A Personal Survival Budget

**Personal survival budget**

| | **£ Week** |
|---|---|
| Mortgage or rent | |
| Rates | |
| Water charges | |
| Gas/oil/coal/Calor gas | |
| Electricity | |
| Telephone (landline) | |
| Telephone (mobile) | |
| Internet connection | |
| House insurance | |
| Contents insurance | |
| Life assurance | |
| Pension | |
| House-keeping | |
| Food | |
| Cleaning | |
| Clothes | |
| Tax | |
| Insurance | |
| Repairs | |
| Fuel | |
| Vehicle costs (percentage not charged to business) | |
| Total carry forward | |

| | **£ Week** |
|---|---|
| **Total brought forward** | |
| Travel expenses | |
| Holidays / Days out | |
| TV license | |
| Newspapers/magazines | |
| Entertainment/ eating out | |
| Subscriptions | |
| Children's pocket money | |
| Presents | |
| Hire purchase payments | |
| TV rental | |
| Court orders | |
| Maintenance payments | |
| Repairs to home | |
| Replacement costs | |
| Credit card bill | |
| Total | |
| Less any other income (benefits, pensions etc | |
| Total necessary to survive | |

## Profit plan

| Month | 1 | 2 | 3 | 4 | 5 | 6 | 7 | 8 | 9 | 10 | 11 | 12 | Total |
|---|---|---|---|---|---|---|---|---|---|---|---|---|---|
| Sales | | | | | | | | | | | | | |
| Income | | | | | | | | | | | | | |
| Total | | | | | | | | | | | | | |
| Expenditure | | | | | | | | | | | | | |
| Purchases | | | | | | | | | | | | | |
| Wages | | | | | | | | | | | | | |
| Drawings | | | | | | | | | | | | | |
| NI and Tax | | | | | | | | | | | | | |
| Vehicle | | | | | | | | | | | | | |
| Rent/rates | | | | | | | | | | | | | |
| Stationary | | | | | | | | | | | | | |
| Advertising | | | | | | | | | | | | | |
| Electricity | | | | | | | | | | | | | |
| Gas | | | | | | | | | | | | | |
| Telephone | | | | | | | | | | | | | |
| Professional | | | | | | | | | | | | | |
| Insurance | | | | | | | | | | | | | |
| Repairs | | | | | | | | | | | | | |
| Interest | | | | | | | | | | | | | |
| Bank charges | | | | | | | | | | | | | |
| Depreciation | | | | | | | | | | | | | |
| Total | | | | | | | | | | | | | |
| Monthly suplus/deficit | | | | | | | | | | | | | |
| Cumulative surplus/deficit | | | | | | | | | | | | | |

# Appendix II
# Risk Assessment Sheet

Risk Assessment sheet for courses run by Rebecca Oaks, Coppice Merchant.

Event: Woodland workshop for Lancaster University.

Details: Research group of up to ten 16–21-year-olds, who will have access to the woods to roam, to attend a pole-lathe workshop.

Location:

Grid reference:

Nearest A & E:

| The Hazard | Worst likely outcome | How likely | Level of risk | Controls |
|---|---|---|---|---|
| Tripping in woodland | Major injury | Possible | Medium | Warn participants of the dangers and ask them to take care. |
| Slipping on limestone pavement | Major injury | Likely | High | Avoid areas of exposed pavement, particularly if the weather is wet. |
| Tree falling | Fatal | Improbable | Med | Avoid the woods altogether in extreme weather. |
| Stick in eye | Major injury | Possible | Med | Warn participants of the risk and ask them to take care. |
| Moving vehicles | Fatal | Possible | High | Vehicles to be kept to a minimum and allocated safe parking space. |
| Use of hand-tools | Major injury | Likely | High | Ensure that all participants are shown how to use tools safely and are supervised by experienced staff on a ratio of 1:5.<br><br>Provide protective gloves and advise on their use. |

Outcomes: Minor injury, Major injury, Fatal. Likelihood: Improbable, Possible, Likely. Level: Low, Medium, High

Signed:

Address:

# Appendix III
# Health and Safety Policy

## HEALTH AND SAFETY POLICY

### *Risk Assessment*

We undertake to make a careful examination of potential hazards involved in the work undertaken, decide on the significance of that hazard and ensure that the risks are minimized through taking satisfactory precautions.

### *Slips Trips and Falls*

We are aware of the hazards associated with the working conditions and take steps to minimize danger from slips, trips and falls by:

- Wearing correct boots and clothing.
- Avoiding extreme conditions (e.g. ice on limestone pavements).
- Not working in conditions of poor visibility.
- Working carefully without rushing when conditions are hazardous.

### *Hazardous Substances*

We are aware of the hazards associated with flammable substances and take steps to minimize risk by:

- Complying with all legal requirements relating to the storage, handling, use and disposal of hazardous substances.
- Keeping hazardous substances in labelled containers.
- Keeping flammable substances away from sources of heat or fire.
- Avoiding contact with skin and using gloves when handling hazardous substances.
- Being aware of the hazardous nature of certain plants and animals.

### *Musculoskeletal Disorders*

We are aware that the nature of our work involves heavy lifting, repetitive and forceful activities and take steps to minimize risk by:

- Undertaking relevant training for ourselves and any employees in safe working techniques.
- Using appropriate lifting equipment when required.

### *Noise*

We will take steps to minimize risks from excessive noise by:

- Wearing the correct protective equipment when operating noisy machinery.

### *Work Equipment and Machinery*

We will take steps to minimize risks associated with work equipment and machinery by:

- Complying with all legal requirements relating to the safe use of machinery and equipment.
- Ensuring that equipment is appropriate for the job.
- Wearing the correct safety equipment and clothing.
- We will ensure that equipment is maintained on a regular basis.
- We will undertake to train any employee in the safe use of any machinery or equipment.
- Adhering to the regulations concerning safe distances between workers when felling or operating machinery.

### *Transport*

We are aware of the hazards presented by vehicles required in undertaking our work and minimize risks by:

- Ensuring care is taken when vehicles are moving about on site.

### *Fire and Explosion*

We are aware of the dangers involved in fire and take steps to minimize risk by:

- Only having fires on agreed sites.
- Only burning suitable material which will not produce dark smoke.
- Taking extreme care when approaching fires; removing tripping hazards.
- Wearing protective clothing, including boots and gloves.
- Avoiding burning in adverse weather conditions i.e. periods of drought or high winds.
- Avoiding conditions which might present a fire hazard on peat i.e. burning on a raised, fire proof platform.

### *First-Aid and Accident Reporting*

We are aware of the general risk of harm to individuals through the nature of our work and take steps to reduce risk by:

- Ensuring that we have a first-aider on site.
- Keeping a first-aid kit on site at all times.
- Ensuring that, wherever possible, we remain in phone communication in case of emergency.
- Carrying the correct insurance that is required.
- Reporting all accidents and incidents in an accident book, which will be kept at:....
- Having a professional approach to leaving a site in a tidy and safe condition.

# Glossary

**Adventitious buds** – produced in the bark below the coppice cut or even on roots just below the ground level.
**Ancient woodland** – a woodland that has existed continuously since at least AD1600 and, therefore, possibly since pre-history.
**Barking** – peeling bark from a tree, normally oak, for use in tanning.
**Boy** – a device for clamping twigs during the process of making a faggot or besom.
**Bender** – a temporary shelter made from using bent poles covered by polythene or canvas sheeting.
**Besom** – 1. A birch broom – the kind that witches use, or are used for quidditch. 2. Description of the kind of birch twigs that are used to make such a birch broom.
**Billet** – a short length of wood, varying from about 8in (20cm) to 36in (91cm), sometimes split.
**Black heart** – discoloration in the centre of a tree – normally ash; does not always signify rot.
**Brash** (also brish or brushwood) – the small twiggy branches resulting from dressing out the tops and side branches from coppice or standards.
**Broache** – another name for thatching spar.
**Bodging** – the art of rustic chair making, often in the wood, and usually making use of a pole-lathe to make the round components.
**Bool** – the hazel rim of a swill basket.
**Burrow** – an area of coppice cut or sold in a season; see also cant, coupe, panel, hagg, sale and spring.
**Butt** – the lowest portion of a trunk, stem or pole.
**Cant** – an area of coppice cut or sold in a season; see also burrow, hagg, sale, panel, spring and coupe.
**Cleft** – to split a segment of wood from a round pole; see also rive.
**Coppice** – underwood trees, which are cut, close to ground level every few years to allow multiple stems to grow again from the stool.
**Coppice-with-standards** – system of coppice management with scattered, single-stemmed trees, such as oak or ash; see also standards.
**Coppicing cycle** – the number of years between cutting of the coppice see rotation.
**Crown** – the living branches of a tree above the main stem.
**Coupe** – an area of coppice cut or sold in a season; see also cant, hagg, sale, panel, spring, fell and burrow.
**Deadwood** – valuable biodiversity resource – can be on the ground or standing.
**Dillaxe** – another name for a froe, especially in Kent.
**Drawn hazel** – to take only the rods you need from a coppice stool, rather than felling the whole stool.
**Drifts** – cut coppice material or brash laid in rows for sorting or disposal; see also windrows.
**Epicormic shoots** – shoots sprouting from dormant or adventitious buds on a tree's main stem.
**Epiphyte** – a plant growing on another without being parasitic.
**Ethers** – the flexible rods that tie the top of a laid hedge together – usually hazel.
**Extraction** – the removal of felled timber from woodland.
**Faggot** – bundle of twigs and small brash tied tightly for use as a firelighter.
**Fell** – an area of coppice cut or sold in a season; see also cant, hagg, sale, panel, spring, coupe and burrow.
**Felling cut** – the cut made from the back of the stem which fells the tree; also known as the back cut.
**Field layer** – the part of the woodland structure containing low growing shrubs, herbaceous plants, grasses and ferns.

**Flush** – 1. An area of ground receiving nutrient-rich water seepage. 2. The first spurt of growth after the winter.
**Formative pruning** – the pruning of branches, usually between 3 and 10 years of planting, in order to improve timber quality.
**Gad** – hazel rod before it is cleft to make a spar or broache.
**Green wood** – freshly felled living wood, still retaining its sap.
**Green wood-work** – the range of crafts that usually entail the use of green wood, such as bodging.
**Ground layer** – the part of the woodland structure that compromises the lowest layer of plant growth; may include flowering plants, grasses, sedges, mosses, liverworts, and fungi; see also field layer.
**Habitat pile** – an artificially made heap of brash and other deadwood specifically constructed to enrich biodiversity.
**Hagg** – an area of coppice cut or sold in a season; see also cant, fell, spring, sale, coupe, panel and burrow.
**Hanger** – a wood growing on the side of a hill; term is used mainly in central southern England.
**Hardwood** – any broad-leaved tree, irrespective of the actual hardness of the wood; see softwood.
**Heartwood** – the inner wood of large branches and trunks that no longer carries sap.
**Hew** – to shape a log with an axe or adze.
**High forest** – woodlands dominated by full-grown trees, often with little shrub layer.
**Layering** – the practice of bending over living shoots and pegging them down to establish a new shrub in a gap – carried out with hazel and sometimes sweet chestnut.
**Leader** – the main top shoot of a tree.
**Lopping** – cutting branches from a tree.
**Maiden** – a single stemmed tree, never coppiced or pollarded. Any tree not grown from a coppice stump; see also standard.
**Motty peg** – the wooden peg that lies at the heart of a charcoal earthburn, around which wood is stacked – removed at the start of the burn.
**Natural regeneration** – trees and shrubs which arise from naturally produced seeds.
**Overstood** – coppice that is still standing beyond its normal rotation.
**Panel** – an area of coppice cut or sold in a season; see also cant, fell, spring, coupe, hagg, sale and burrow.
**Plantation woodland** – where the majority of trees have been planted.
**Pimp** – a southern English term for a tightly tied group of faggots used for fire-lighting.
**Pole** – a coppice shoot of more than 50mm (2in) diameter.
**Pole-lathe** – a primitive kind of foot-operated lathe traditionally used for turning chair components (see bodging), constructed by bending a sapling over to provide the spring required to turn the lathe.
**Pollard** – a tree that is cut at between 2 and 4m (6–12ft) above ground level to produce a crop of poles or branches.
**Prince** – a single-stemmed tree, never coppiced or pollarded. Any tree not grown from a coppice stump; see also standard.
**Prog** – a stout, forked pole used for pushing and levering trees during felling or for turning the remains of a fire.
**Provenance** – the place of origin of a tree stock, which remains the same no matter where later generations of the tree are raised.
**Recent woodland** – woodland which has grown up since AD1600, on land that had previously been cleared, or was previously not wooded; see also secondary woodland.
**Reserve** – a single-stemmed tree, never coppiced or pollarded. Any tree not grown from a coppice stump; see also standard.
**Ride** – a wide woodland road, normally unsurfaced.
**Rive** – to split or cleave a piece of round wood.
**Rod** – small flexible underwood stem of less than 50mm (2in) diameter.
**Rotation** – length of time between the cutting of a coppice coupe; see also coppice cycle.
**Roundwood** – wood of small diameter often used for fencing stakes.
**Ruderal** – describing plants that easily invade bare ground and which disappear again after a few years.
**Sale** – 1. An area of coppice cut or sold in a season; see also cant, fell, panel, spring, coupe, hagg and burrow. 2. The upright in a hazel

hurdle (sometimes spelt zale).
**Sapwood** – wood that carries the sap within a tree stem. This may be all the wood in a young stem or the outermost layer in an older, larger trunk or branch.
**Secondary woodland** – woodland growing on a site that was formerly not woodland. It could be ancient, if it grew up before 1600; see also recent woodland.
**Semi-natural woodland** – in ancient sites, wood made up of native species, where their presence is natural rather than planted. More recently, woods which have originated mainly by regeneration.
**Set** – a large unrooted cutting, usually willow or poplar.
**Short rotation coppice** – coppice grown on a short rotation, normally of up to about 3 years; usually consists of willow or poplar that is chipped and used in wood-burning boilers.
**Short rotation forestry** – trees (only sometimes coppiced), grown on a short rotation of up to about 10 years; would normally consist of ash, alder, birch or eucalyptus.
**Shrub layer** – the part of the woodland structure that includes shrubs and young trees.
**Singling** – retaining one stem on a coppice stool and allowing it to grow into a standard tree; sometimes done to a whole coppice wood to create high forest.
**Softwood** – the timber of a coniferous tree, irrespective of the hardness of the timber.
**Spar** – hazel rod cleft and twisted used for pinning thatch on a roof.
**Spelk** – another name for swill basket, but also describes the oak laths used to weave the basket from the Norse meaning splinter.
**Spring** – 1. The first new coppice regrowth. 2. An area of coppice cut or sold in a season; see also cant, fell, panel, sale, coupe, hagg and burrow.
**Staddle** – a single-stemmed tree, never coppiced or pollarded. Any tree not grown from a coppice stump. Perhaps especially a young tree yet to become of value as timber; see also standard.
**Standard** – a single-stemmed tree, never coppiced or pollarded. Any tree not grown from a coppice stump; see also maiden, prince, reserve, staddle, standil, store, teller and waverer.
**Standil** (also spelt standell, standle, standrell) – a single-stemmed tree, never coppiced or pollarded. Any tree not grown from a coppice stump; see also standard.
**Stool** – the base of a coppiced tree from which new shoots emerge.
**Stored coppice** – coppice that has grown beyond its normal rotation, but which is still capable of being brought back into rotation.
**Stores** – a single-stemmed tree, never coppiced or pollarded. Any tree not grown from a coppice stump; see also standard.
**Sucker** – shoots growing from the roots of an older tree.
**Swill** (or spelk) – a basket made from woven boiled oak sapwood strips, now only made in the Lake District.
**Teller** – a single-stemmed tree, never coppiced or pollarded. Any tree not grown from a coppice stump; see also standard.
**Timber** – tree trunk suitable for making beams or sawing into planks, normally derived from standards.
**Twilly hole** – a hole designed into a woven hazel hurdle to enable hurdles to be carried over the shoulder on a pole.
**Underwood** – coppiced shrub layer growing under standard or timber trees.
**Up-and-down wood** – small-dimension firewood, so-called because it doesn't last long on a fire or in a stove, meaning that you are always getting up and down to feed the fire.
**Waverer** – a single-stemmed tree, never coppiced or pollarded. Any tree not grown from a coppice stump; see also standard.
**Wildwood** – term used to describe ancient forest, untouched by human activity; doesn't exist in the UK.
**Windrow** – a linear pile of brash.
**Wood** – 1. The part of the stem, inside the cambium, that supports the tree, carries water to the crown and stores reserves of food over the winter period. 2. Sometimes used (especially historically) to denote poles and branches, which are distinct as a product to timber.

# Bibliography


Bright, P., Morris, Pat. and Mitchell-Jones, T., *The Dormouse Conservation Handbook* (Natural England 2nd Edition, 2006).

Buckley, G.P. (ed.), *Ecology and Management of Coppice Woodlands* (Chapman and Hall, 1992).

Collins, E.J.T., *Crafts in the English Countryside* (Countryside Agency Publications, 2004).

Edlin, H., *Woodland Crafts in Britain* (David and Charles, 1973).

Emrich, W., *Handbook of Charcoal Making* (Kluwer Academic Publishers Group, 1995).

Evans, J., Badgers, *Beeches and Blisters* (Patula Books, 2006).

Forestry Commission, *Utilization of Hazel Coppice* (FC Bulletin No. 27, 1956).

Forestry Commission, Forestry Advice Note 2 *Managing Deer in the Countryside* (Forestry Commission, 1995).

Harmer, R., *Restoration of Neglected Hazel Coppice* (Forestry Commission Information Note, 2004).

Harmer, R. and Howe, J., *The Silviculture and Management of Coppice Woodlands* (Forestry Commission, 2003).

Harmer, R. and Robertson, M., *Management of Standards in Hazel Coppice*, Quarterly Journal of Forestry, Vol. 96, No. 4, pp. 259–264.

Howe, J., *Hazel Coppice – Past, Present and Future* (Second edition Hampshire County Council, 1995).

Kelly, D., *Charcoal and Charcoal Burning* (Shire Publications, 1996).

Land Use Consultants, *Kent Wood-Lotting* (2007).

Marren, P., *Britain's Ancient Woodland Heritage* (The Nature Conservancy Council, 1990).

Miles, A., *Silva the Tree in Britain* (Felix Dennis, 1999).

Rackham, O., *Trees and Woodland in the British Landscape* (1986).

Rackham, O., *Ancient Woodland* (Castlepoint Press, 2003).

Rackham, O., *Woodlands* (Collins New Naturalist, 2006).

Shepley, A., (ed.), *21st Century Coppice* (Wood Education Programme Trust and Coppice Association North West, 2007).

Smart, R. and Wellings, R. (ed.), Worcestershire *Woodin'* (Small Woods Association, 2009).

Spooner, B. and Roberts, P., *Fungi* (Collins, 2005).

Symes, N. and Currie, F., *Woodland Management for Birds* (RSPB, 2005).

UKWAS – *UK Woodland Assurance Standard*, second edition (UKWAS, amended 2008).

# Further Reading


Abbott, M., *Green Woodwork* (Guild of Master Craftsman Publications Ltd, 1991).

Abbott, M., *Living Wood* (Living Wood Books, 2002).

Armstrong, L,. *Woodcolliers and Charcoal Burning* (Coach Publishing House Ltd and The Weald and Downland Open Air Museum, 1978).

Evans, J., *A Wood of our Own* (Oxford University Press, 1995).

Jenkins, J.G., *Traditional Country Craftsmen* (Routledge and Kegan Paul Ltd, 1978).

King, P., *The Complete Yurt Handbook* (Eco-Logic Books 2001).

Lambert, F., *Tools and Devices For Coppice Crafts* (Young Farmers' Club, Booklet 31, 1979).

Law, B., *The Woodland Way* (Permanent Publications, 2001).

Mitchell, A., *Trees of Britain and Northern Europe* (Collins, 1982).

Olsen, L.-H., Suneson, J. and Pederson, B.V., *Small Woodland Creatures* (Oxford University Press, 2001).

Peterken, G.F., *Woodland Conservation and Management* (Chapman and Hall, 1991).

Shepley, A. (ed.), Bill Hogarth MBE: *Coppice Merchant* (Wood Education Programme Trust, 2001).

Tabor, R., *Traditional Woodland* Crafts (Batsford, 1994).

Tolman, T., *Photographic Guide to the Butterflies of Britain and Europe* (Oxford University Press, 2001).

Sinclair, G. and Kenny, K., *Growing Your Own Beanpoles* (undated).

Starr, C. *Woodland Management – A Practical Guide* (The Crowood Press, 2005)

Waring, P. and Townsend, M., *Field Guide to Moths of Great Britain and Ireland* (British Wildlife Publishing, 2009).

# Useful Addresses

Association of Pole-lathe Turners and Greenwood Workers
David Reeve
Little Malt House
Ockham Road North
East Horsley, Surrey
KT24 6PU
www.bodgers.org.uk

British Deer Society
Fordingbridge
Hampshire SP6 1EF
www.bds.org.uk

British Trust for Conservation Volunteers (BTCV)
Sedum House
Mallard Way
Doncaster
DN4 8DB
www.btcv.org

British Trust for Ornithology
The Nunnery
Thetford
Norfolk
IP24 2PL
www.bto.org.uk

Confederation of Forest Industries
59 George St
Edinburgh
EH2 2JG
Tel: 0131 240 1410
www.confor.org.uk

Coppice Association North West (CANW)
C/o Cumbria Woodlands
Staveley Mill Yard
Staveley in Kendal
Cumbria
LA8 9LR
www.coppicenorthwest.org.uk

Countryside Council for Wales
Maes y Ffynnon
Penrhosgarnedd
Bangor
Gwynedd
LL57 2DW
Tel: 0845 1306 229

The Deer Initiative
Head Office
The Carriage House
Brynkinalt Business Centre
Chirk
Wrexham
LL14 5NS
Tel: 0845 872 4956
www.thedeerinitiative.co.uk

DEFRA
Customer Contact Unit
Eastbury House
30–34 Albert Embankment
London
SE1 7TL
Helpline: 08459 33 55 77
helpline@defra.gsi.gov.uk
www.defra.gsi.gov.uk

Deer Commission for Scotland (DCS)
Great Glen House,
Leachkin Road,
Inverness,
IV3 8NW
Tel: 01463 725000
www.dcs.gov.uk

Forest Research
Alice Holt Lodge
Farnham
Surrey GU10 4LH
01420 22255

Forest Research Northern Research Station
Roslin
Midlothian EH25 9ST
0131 445 2176
www.forestresearch.co.uk

Forest Research
Forestry Commission Wales
Welsh Assembly Government
Rhodfa Padarn
Llanbadarn Fawr
Aberystwyth
Ceredigion SY23 3UR
Tel: 0300 068 0300

Forest Stewardship Council (FSC)
11–13 Great Oak Street
Llanidloes
SY18 6BU
Tel: 01686 413916
www.fsc-uk.org

Forestry Commission HQ
Public Enquiries
231 Corstorphine Road
Edinburgh
EH14 5NE
Tel: 0845 3673787
www.forestry.gov.uk

Forestry Commission Publications Section
PO Box 501
Leicester
LE94 0AA
Tel: 0844 991 6500
forestry@mrm.co.uk

Forestry Contracting Association (FCA)
www.fcauk.com

Game Conservancy Trust
Burgate Manor
Fordingbridge
Hampshire
SP6 1EF
Tel: 01425 652381

Health and Safety Executive (HSE)
Head Office
Health and Safety Executive
(1G) Redgrave Court
Merton Road
Bootle
Merseyside
L20 7HS
Tel. info line: 0845 345 0055
www.hse.gov.uk

Institute of Chartered Foresters
59 George Street
Edinburgh EH2 2JG
Phone: 0131 240 1425
www.charteredforesters.org

Lantra Awards
Lantra House
Stoneleigh Park
nr Coventry
Warwickshire CV8 2LG
Telephone: 02476 419 703
www.lantra-awards.co.uk

The Mammal Society
3 The Carronades,
New Road,
Southampton
SO14 0AA
Tel. 0238 0237874
www.mammal.org.uk

City and Guilds – National Proficiency Test Council (NPTC)
Building 500
Abbey Park
Stareton
Warwickshire
CV8 2LY
Tel: 024 7685 7300
www.nptc.org.uk

National School of Forestry
University of Cumbria
Newton Rigg
Penrith
CA11 0AH
Tel: 01768 893400
www.forestry.org.uk

The National Trust
PO Box 39
Warrington WA5 7WD
Tel: 0844 800 1895
www.nationaltrust.org.uk

Natural England
1 East Parade
Sheffield, S1 2ET
Tel: 0845 600 3078
www.naturalengland.org.uk

Northern Ireland Forest Service
Dundonald House
Upper Newtownards Road
Belfast
BT4 3SB
Tel: 02890 524480
www.forestserviceni.gov.uk

Reforesting Scotland
58 Shandwick Place
Edinburgh
EH2 4RT
Tel: 0131 2202500
www.reforestingscotland.org

Royal Forestry Society on England Wales and Northern Ireland
102 High Street
Tring
Hertfordshire
HP23 4AF
Tel: 01442 822028
www.rfs.org.uk

Royal Scottish Forestry Society
Potholm
Langholm
Dumfriesshire
DG13 0NE
Tel: 01387 383845
www.rsfs.org.uk

Royal Society for the Protection of Birds (RSPB)
The Lodge
Potton Road
Sandy
Bedfordshire
SG19 2DL
Tel: 01767 680551
www.rspb.org.uk

Royal Society of Wildlife Trusts
The Kiln
Waterside
Mather Road
Newark
Nottinghamshire
NG24 1W
Tel: 01636 677711
www.wildlifetrusts.org

Scottish Natural Heritage
Great Glen House,
Leachkin Road
Inverness
IV3 8NW
Tel:01463 725000
Fax: 01463 725067
www.snh.org.uk

1) Small Woods Association
2) Green Wood Centre
3) Woodlands Initiatives Register
4) National Coppice Apprenticeship
Green Wood Centre
Station Road
Coalbrookdale
Telford
TF8 7DR
Tel: 01952 432769
www.smallwoods.org.uk

Sustainability Centre
Mercury Park
East Meon
Petersfield
Hampshire
GU32 1HR
01730 823 166
www.sustainability-centre.org

Tree Council
71 Newcomen Street
London
SE1 1YT
Tel: 020 7407 9992
www.treecouncil.org.uk

Tree Advice Trust
Alice Holt Lodge
Wrecclesham
Farnham
Surrey
GU10 4LH
Tel. helpline: 09065 161147
www.treehelp.info

The Vincent Wildlife Trust
3 and 4 Bronsil Courtyard
Eastnor
Ledbury
Herefordshire
HR8 1EP
Tel: 01531 636441

Woodland Heritage
PO Box 168
Haslemere
Surrey
GU27 1XQ
Tel: 01428 652159
www.woodlandheritage.org

The Woodland Trust
Autumn Park
Dysart Road
Grantham
Lincolnshire
NG31 7DD
Tel: 01476 581 111
www.woodlandtrust.org.uk

### *Suppliers*

Lakeland Coppice Products (Tools)
www.lakelandcoppiceproducts.co.uk

Selway Packaging
(Charcoal bags)
www.selway.co.uk

Woodland Craft Supplies Ltd
www.woodlandcraftsupplies.co.uk

The Woodsmiths Store
www.woodsmithstore.co.uk

# Index